Cloud and Edge Networking

SCIENCES

Networks and Communications, Field Director – Guy Pujolle
Cloud Networking, Subject Head – Kamel Haddadou

Cloud and Edge Networking

Kamel Haddadou
Guy Pujolle

WILEY

First published 2023 in Great Britain and the United States by ISTE Ltd and John Wiley & Sons, Inc.

Apart from any fair dealing for the purposes of research or private study, or criticism or review, as permitted under the Copyright, Designs and Patents Act 1988, this publication may only be reproduced, stored or transmitted, in any form or by any means, with the prior permission in writing of the publishers, or in the case of reprographic reproduction in accordance with the terms and licenses issued by the CLA. Enquiries concerning reproduction outside these terms should be sent to the publishers at the undermentioned address:

ISTE Ltd
27-37 St George's Road
London SW19 4EU
UK
www.iste.co.uk

John Wiley & Sons, Inc.
111 River Street
Hoboken, NJ 07030
USA
www.wiley.com

© ISTE Ltd 2023
The rights of Kamel Haddadou and Guy Pujolle to be identified as the authors of this work have been asserted by them in accordance with the Copyright, Designs and Patents Act 1988.

Any opinions, findings, and conclusions or recommendations expressed in this material are those of the author(s), contributor(s) or editor(s) and do not necessarily reflect the views of ISTE Group.

Library of Congress Control Number: 2022950912

British Library Cataloguing-in-Publication Data
A CIP record for this book is available from the British Library
ISBN 978-1-78945-128-3

ERC code:
PE7 Systems and Communication Engineering
 PE7_8 Networks (communication networks, sensor networks, networks of robots, etc.)

Contents

Preface xi

Chapter 1. Introduction to Edge and Cloud Networking 1

    1.1. Introduction to the digital infrastructure 1
    1.2. Cloud services 7
    1.3. Cloud Networking 9
    1.4. Network Functions Virtualization 14
    1.5. Conclusion 16
    1.6. References 16

Chapter 2. The Cloud Continuum 19

    2.1. Cloud Continuum levels 19
    2.2. Cloud Continuum Networks 22
    2.3. The Cloud Continuum and the digitization of companies 23
    2.4. Example of digital infrastructure 25
    2.5. Conclusion 28
    2.6. References 28

Chapter 3. Digital Infrastructure Architecture 31

    3.1. The evolution of enterprise information system architectures 31
    3.2. The Open Infrastructure Foundation architecture 36
    3.3. The Cloud Native Computing Foundation architecture 42

3.4. Gaia-X. 49
3.5. Conclusion 54
3.6. References 54

Chapter 4. Open-Source Architectures for Edge and Cloud Networking 57

4.1. Organizations and the main open sources 57
4.2. The main open-source projects 57
4.3. Conclusion 69
4.4. References 70

Chapter 5. Software-Defined Networking (SDN) 73

5.1. Introduction to Software-Defined Networking 73
5.2. ONF architecture 74
5.3. Southbound interfaces and controllers 80
5.4. The northbound interface and the application plan. 82
5.5. Conclusion 84
5.6. References 85

Chapter 6. Edge and Cloud Networking Commercial Products . . . 87

6.1. Introduction to SDN products 87
6.2. Fabric control. 87
    6.2.1. NSX from VMware. 89
    6.2.2. Cisco Application Centric Infrastructure 92
    6.2.3. OpenContrail and Juniper 94
    6.2.4. Nokia SDN Architecture. 95
6.3. Software-Defined Wide Area Network 96
    6.3.1. The basics of SD-WAN 96
    6.3.2. SD-WAN 2.0 101
    6.3.3. SD-Branch 102
6.4. Secure Access Service Edge. 103
6.5. Virtual Customer Premises Equipment 105
6.6. vWi-Fi. 107
6.7. Virtual Radio Access Network 109
6.8. Virtual Evolved Packet Core and virtual 5GCore 110
6.9. Conclusion 111
6.10. References. 111

Chapter 7. OpenFlow, P4, Opflex and I2RS 113

7.1. OpenFlow signaling. 113
7.2. P4. 120
7.3. OpFlex. 121
7.4. I2RS 122
7.5. Conclusion 123
7.6. References 124

Chapter 8. Edge and Cloud Networking Operators 127

8.1. Edge Networking in 5G architecture 127
8.2. Cloud RAN. 130
8.3. Cloud Networking at the heart of 5G. 132
8.4. The Cloud and the new Ethernet and Wi-Fi generations 134
8.5. Enterprise 5G Edge Networks. 136
8.6. Conclusion 138
8.7. References 138

Chapter 9. Cloud Networking Protocols. 141

9.1. Low-level protocols. 142
   9.1.1. Radio over Fiber 143
   9.1.2. Ethernet over Fiber 144
9.2. Virtual extensible LAN. 144
9.3. Network Virtualization using Generic Routing Encapsulation. 146
9.4. Ethernet MEF 146
9.5. Ethernet Carrier Grade 147
9.6. Transparent Interconnection of Lots of Links. 150
9.7. Locator/Identifier Separation Protocol 152
9.8. Conclusion 153
9.9. References 153

Chapter 10. Edge and Cloud Networking in the IoT. 155

10.1. Internet of Things networks 156
10.2. Low Power Wide Area Networks. 158
10.3. PAN and LAN networks for the IoT 162
10.4. Telecommunications operator networks for the IoT 166

10.5. Platform for the IoT 169
10.6. Conclusion 178
10.7. References. 178

Chapter 11. Cloud Continuum in Vehicular Networks 181

11.1. ETSI ITS-G5 183
11.2. 5G standardization 185
    11.2.1. 5G vehicular networks 185
    11.2.2. C-V2X technology overview 187
11.3. Visible light communication 189
11.4. The architecture of vehicular networks. 190
11.5. Conclusion 193
11.6. References. 193

Chapter 12. The Cloud Continuum and Industry 4.0 199

12.1. The features needed to achieve Industry 4.0. 201
12.2. Technical specifications for 5G 203
12.3. Cloud and Edge for Industry 4.0. 205
12.4. Conclusion 207
12.5. References. 208

Chapter 13. AI for Cloud and Edge Networking 211

13.1. The knowledge plane 211
13.2. Artificial intelligence and Software-Defined Networking. 214
13.3. AI and Cloud Networking management 217
13.4. AI through digital twins. 218
13.5. Conclusion 221
13.6. References. 223

Chapter 14. Cloud and Edge Networking Security 229

14.1. The Security Cloud 229
14.2. SIM-based security 230
14.3. Blockchain and Cloud. 233
14.4. Cloud Networking security 234

14.5. Edge Networking security 241
  14.5.1. Security of 5G MEC 241
  14.5.2. Threats to Network Functions Virtualization 242
  14.5.3. Fog security 243
  14.5.4. Protection of intelligent processes in the Edge 244
  14.5.5. Client security through the use of HSM 245
14.6. Conclusion 246
14.7. References 247

Chapter 15. Accelerators 253

15.1. The DPDK accelerator 254
15.2. The FD.io accelerator 258
15.3. Hardware virtualization 260
15.4. Conclusion 263
15.5. References 263

Chapter 16. The Future of Edge and Cloud Networking 267

16.1. 5G continuity 269
16.2. Fully distributed networks 272
16.3. Cloud Continuum-based networks 275
16.4. Edge and Cloud properties 276
16.5. Conclusion 278
16.6. References 278

Conclusion 283

List of Authors 285

Index 287

Preface

Cloud Networking is revolutionizing the field of networks by proposing a complete paradigm shift. The Cloud imposes a completely different vision of the Internet with centralized controls and a virtualization of all the functions necessary to the life of a network, the latter consisting of the replacement of hardware equipment by software equipment which runs in the Cloud. As a result, physical networks are replaced by logical networks that run in data centers that are more or less remote from the network nodes themselves.

The broad categories of Clouds, with data centers ranging from infinitely large to infinitely small, are examined and described in detail. Data centers within 10 km form the Edge and support real-time applications with latencies of less than 1 ms. This set of data centers takes the shape of the Cloud Continuum, which becomes the core environment for Cloud Networking. The logical devices that replace the physical devices must be urbanized in this environment, that is, positioned in the Cloud Continuum in the optimal location for the network to work best.

This book then introduces the digital infrastructure of the 2020s, which is composed of three elements: an antenna that collects the signals from users, an optical fiber that carries the signal and the data center that receives this signal, processes it and performs all the functions requested by the user. All material devices and intermediary equipment are recreated as virtual machines. Thus, it is possible to add many functions such as control, management, security, intelligence, automation and so on.

Another feature is the use of open source software, which seems to be self-evident since the whole point of this new generation of Cloud Networking is to be able to reduce costs despite the increase in user throughput that doubles every

Cloud and Edge Networking,
by Kamel HADDADOU and Guy PUJOLLE. © ISTE Ltd 2023.

year. The agility and flexibility of this approach makes it an incomparable solution that is widely introduced in this book.

This new generation of software-defined networking technology translates into a number of products, described in detail, including SD-WAN, which constitutes the main requirement of large enterprises along with vCPE (virtual Customer Premises Equipment) and data center access fabrics. The impact of Cloud Networking is equally important in carrier networks and network providers. It is the foundation of 5G and even forms the revolutionary aspect of this generation. Indeed, what is typically called 5G relates to the radio part of this technology, but that is not the most important part. The revolutionary part is the MEC (Multi-access Edge Computing) data centers that sit on the edge of the Edge with response times that support a whole new set of real-time services such as vehicular network automation, Industry 4.0, touch networks and so on.

SDN has a special role in Cloud Networking, providing automated control of the digital infrastructure through centralization. However, this solution has not yet fully imposed itself due to its overly disruptive aspect, simultaneously providing automated control of a new generation of equipment through the migration of intelligence to the central controller but also a level of centralization that can seem too exacerbated.

Finally, we introduce in this book a whole set of important points like security, reliability, intelligence and acceleration. It ends with a vision of what Cloud Networking could become in the future, especially with 6G. Indeed, a return to hardware is more than likely to both improve performance and consume much less energy.

We hope to fulfill the expectations of all those interested in Cloud Networking with a relatively high-level vision of all the elements necessary to fully understand the path of this technology toward 6G.

Kamel HADDADOU and Guy PUJOLLE

July 2023

1
Introduction to Edge and Cloud Networking

1.1. Introduction to digital infrastructure

For the next 10 years, digital infrastructure in Cloud Networking will establish itself as the basic standard. This standard has been adopted by all network and telecommunications equipment manufacturers. It consists of four elements: the terminal equipment, an antenna, an optical fiber and a data center. To understand the reasons that led to this architecture, we must start with the basic element: virtualization.

The virtualization process is described in Figure 1.1. This process is a result of moving from a physical machine to a logical machine. The first step is to write code that does exactly the same thing as the physical machine. Assuming the physical machine is a router, the virtual router code must perform the same routing and send the incoming packet processed by the logical code on the same outgoing line as the physical machine would.

The next step is to compare the performance of the physical machine and the logical machine by running it on the processor of the physical machine. Without accelerator hardware such as ASICs (application-specific integrated circuits) or Field-Programmable Gate Array (FPGAs), performance will easily drop by a factor of at least 10 and possibly as much as 100. If we assume this loss by a factor of 20, it would take a processor 20 times more powerful to achieve the same performance, which is not a problem with data center power. However, since energy consumption is very roughly proportional to processor power, it jumps to a high level.

The next step is to try to minimize the energy expenditure. To do this, the processor of the physical machine supporting the logical machine must be occupied as close to 100% as possible. As this is not really possible, we must try to stay around 80%. To achieve this, a sufficient number of virtual machines must be multiplexed to achieve a very good CPU utilization.

Figure 1.1. *The virtualization process. For a color version of this figure, see www.iste.co.uk/haddadou/edge.zip*

The solution is to group virtual machines so that there are exactly the right number of them. If demand is too high, virtual machines must be migrated to other servers and vice versa to maintain high CPU utilization. We can also see from Figure 1.1 that data center utilization is the solution since the many servers are either put into sleep mode if not in use or they run at high utilization. Optimization of energy consumption is therefore achieved by migrating virtual machines so that all servers not in standby mode are heavily used. Virtual machine migrations, that is, the movement of virtual machines from one server to another, are in the vast majority of cases carried out in the same data center and much more rarely between separate data centers.

Figure 1.2 shows a data center with its virtual machines. As shown, there are continuous migrations to optimize operation. We also need to be able to give the virtual machines the power they need to perform the requested task. To do this, we need an orchestrator of the data center resources that are allocated to the virtual machines.

This software virtualization should be replaced gradually by hardware virtualization because of reconfigurable processors, but it will take many years before this new generation arrives, which will consume much less energy and greatly increase performance.

Introduction to Edge and Cloud Networking 3

Figure 1.2. *A data center and its virtual machines. For a color version of this figure, see www.iste.co.uk/haddadou/edge.zip*

The question arises as to which physical elements can be virtualized and which cannot be virtualized. In fact, it is better to look at the second part of the question since everything is virtualizable except for three elements: the sensors, the wireless communication cards and wired communication cards. Sensors are not virtualizable because they have to capture something, which cannot be done by a code. For example, we cannot measure the temperature in a room by writing a code. In the same way, we cannot capture an electromagnetic signal with a code, nor can we always send a light in an optical fiber by a code. Otherwise, everything is virtualizable: a Wi-Fi box, a firewall, a key, a switch, etc.

Cloud Networking is precisely the network solution that uses the digital infrastructure that was described at the beginning of this chapter, that is, based on four elements: the terminal equipment, the antenna, the optical fiber and the data center. We will start by describing a few types of Clouds and their importance.

The Cloud is above all a mechanism that consists of grouping the resources of a company in the Internet rather than having them directly in the company, in order to share them with other users and benefit from a strong multiplexing of the resources and therefore a reduced cost. Cloud providers also benefit from multiplexing by selling shared resources to users who may be located on different continents.

In the early 2000s, the utilization of hardware, software and personnel resources was not optimized, since these resources were heavily used only during peak hours and hardly at all at night. Average utilization calculations showed that resources were used at less than 20%. By connecting several companies to the same common resources at different peak times, it is possible to achieve utilization rates of around 80% without increasing the resources.

Figure 1.3. *Virtualizable and non-virtualizable devices. For a color version of this figure, see www.iste.co.uk/haddadou/edge.zip*

The problem that immediately arose concerned the data of companies that are in a public Cloud and are therefore often at the mercy of attackers or states requesting information from their providers for cybersecurity reasons.

Private Clouds have been democratized to take this issue into account and have become the majority. The data is often installed on several private clouds within the framework of either large companies with several sites or companies with a single site but independent departments.

Today, there are different types of Clouds that have become more complex to accommodate new diversification and availability parameters needed by businesses.

The first type concerns distributed Clouds. As shown in Figure 1.4, these are different types of Clouds offered by the same provider: public, private, close to the user, on the Edge, or in the core of the Internet network but much further from the user, which we will call the Core Cloud.

The Edge (data centers on the edge of the core network) or Cloud (data centers inside the core network) provider can offer several types of services that we will detail later: an infrastructure, a platform or an application software.

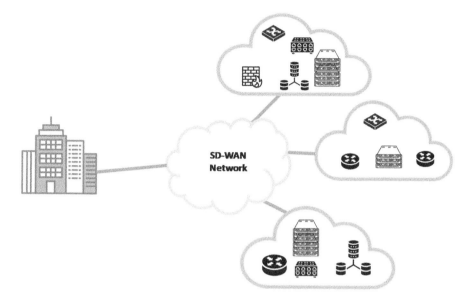

Figure 1.4. *A distributed Cloud. For a color version of this figure, see www.iste.co.uk/haddadou/edge.zip*

Another widely used term is Hybrid Cloud, in which the data centers can be both private and public but can come from different Cloud providers. The hybrid Cloud is therefore a solution that combines a private Cloud, one or more public Cloud services and often proprietary software that enables communication between each service. By opting for a hybrid Cloud strategy, companies gain greater flexibility by shifting loads between different Cloud providers as needs change, and costs can vary rapidly.

An illustration of this type of Cloud is provided in Figure 1.5, where we see the connection of the two Clouds realized by access to multiple applications carried by the public or private part.

Other types have also been defined, such as multi-Clouds, which bring together several providers to support all the services requested by companies and which allow better availability in the event of overload of one of the Clouds in the environment. These multi-Clouds bring together both public and private Clouds and different types of services, platforms, infrastructure and applications.

Finally, the term omni-Cloud is the most general to take into account the multitude of possibilities of associations and structures of Clouds.

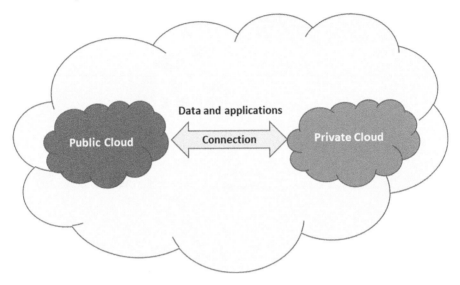

Figure 1.5. *A hybrid Cloud. For a color version of this figure, see www.iste.co.uk/haddadou/edge.zip*

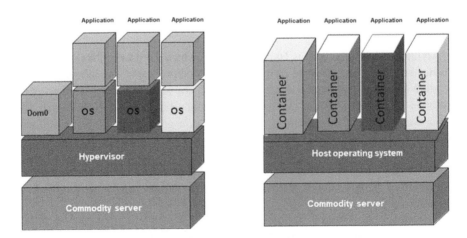

Figure 1.6. *Hypervision and containerization. For a color version of this figure, see www.iste.co.uk/haddadou/edge.zip*

In Figure 1.6, we describe the internal architecture of servers inside a data center. There are two main possibilities: hypervisor and containerization. The first is the older one, it concerns the support of virtual machines as it was originally conceived. The second solution is gradually replacing the first with a simpler, less expensive and more flexible architecture.

Hypervision consists of using a hypervisor on a standard physical machine (commodity) which is a software able to support several virtual machines simultaneously through one or several operating systems (OSs). The hypervisor supports domains formed by an operating system and the virtual machine running on it. The Domain 0 or Dom0 is specialized in processing the I/O of the other domains on the base physical machine.

There are different types of hypervisors. Paravirtualization requires that the operating systems be slightly modified so that all the processing requested by the virtual machine can be done natively on the basic physical machine. On the contrary, the second solution is to accept the operating systems without modification but with the introduction, above the hypervisor, of an emulation software able to adapt the execution of certain functions to the underlying physical machine.

Containerization is gradually replacing hypervisor with a division of services into microservices that each run in a container. In this case, a single operating system is used that supports containers that are isolated from each other to avoid "jumping", allowing a user to move from one container to another. Each microservice runs in a container and application interfaces between microservices allow the service itself to be realized. We will study this microservices technology in more detail in the following section, as it allows for simpler updates without completely stopping the service, and also allows for simplified development of services.

This microservices technology is itself beginning to be replaced by a function-based solution that we will study in detail in the following section, which consists of building services with a succession of functions. This last solution is called serverless to indicate that the programmer who develops a function-based service is no longer at all aware that there are underlying servers.

1.2. Cloud services

Figure 1.7 explains the three major types of Clouds that are complemented by two new Clouds that fall between the solutions shown in Figure 1.7.

The first protocol stack on the left represents the case where a company has all the resources to run its information system. These resources include the network, storage and computing elements on hardware servers. To this must be added the virtualization and the operating system that support the company's data and applications.

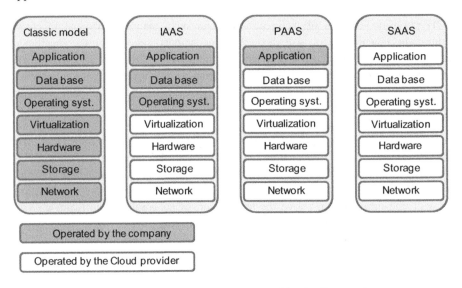

Figure 1.7. *The three main types of Clouds. For a color version of this figure, see www.iste.co.uk/haddadou/edge.zip*

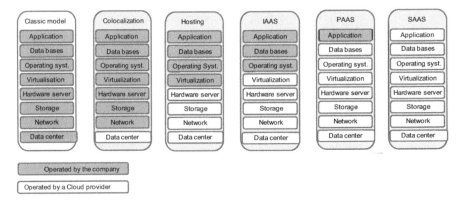

Figure 1.8. *The first service architectures. For a color version of this figure, see www.iste.co.uk/haddadou/edge.zip*

Introduction to Edge and Cloud Networking 9

When using an IaaS (Infrastructure as a Service) provider, the provider delivers the lower layers corresponding to the network, storage and computation with the virtualization environment and the company takes care of the upper layers. In PaaS (Platform as a Service), the provider takes care of everything except for the application. Finally, for SaaS, the provider offers all the layers including the applications. An example of the latter is Microsoft's Office 365 service. SaaS represents approximately 50% of installed Clouds, and the other two share the rest.

We show in Figure 1.8 more classical cases that have been used for a long time and are gradually being replaced by the three cases we have described above. These are mainly colocation and hosting.

Now that we have introduced virtualization and the Cloud, we can move into Cloud Networking.

1.3. Cloud Networking

Cloud Networking is a group of network technologies that are based on the Cloud and thus on virtualization. The digital infrastructure is the basis: all the physical machines, all the services of the digital infrastructure and the application services form the Cloud Networking. First, let us define the virtual networks that form the basis of Cloud Networking.

Figure 1.9 shows a set of virtual networks that are made up of virtual machines, routers, switches, firewalls, authentication servers, etc. to perform all the functions necessary for a network to function properly. Each virtual network is made up of its own virtual machines which can be very different from each other. One network can be made up of IPv6 routers, another of Ethernet switches, a third of LSRs (Label Switch Routers) found in MPLS networks and finally a fourth that has very specific and proprietary equipment. These networks use the same data centers and cable infrastructures or microwave links between the virtual equipment. The number of virtual networks that can coexist on the digital infrastructure depends on the will of the environment manager and the traffic on each network. These virtual networks are called slices, and we will mainly use this word in 5G core networks. These different virtual networks must be independent and isolated from each other to prevent an attack from spreading from one network to another.

Figure 1.10 shows a virtual network built on data centers that become the network rooms of the digital infrastructure. In this figure, apart from the routers or switches, the equipment is not virtualized, such as the Internet boxes at the users' premises or intermediate boxes such as the DPI (Deep Packet Inspection) or the firewall or an authentication server.

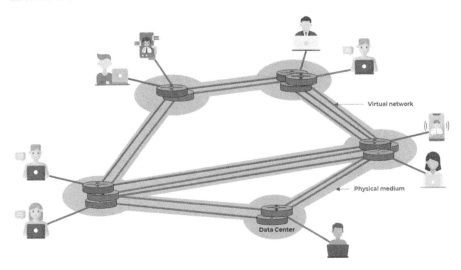

Figure 1.9. *A set of virtual networks. For a color version of this figure, see www.iste.co.uk/haddadou/edge.zip*

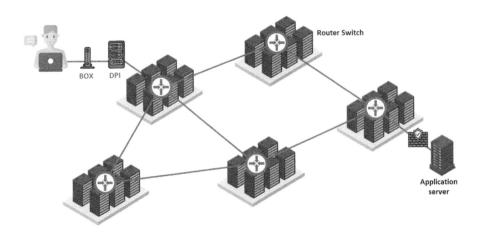

Figure 1.10. *A virtual network with non-virtualized boxes. For a color version of this figure, see www.iste.co.uk/haddadou/edge.zip*

Figure 1.11 introduces slicing, that is, as we have seen, the presence of several virtual networks on the same physical infrastructure. These networks may belong to different operators, each with its own packet transfer technology and specific functions. Slicing is most notably introduced in the 5G core network, the network that interconnects antennas to allow the transfer of information from one region to another. However, the slices are defined from end to end, so they continue on the access and radio parts as we will detail in Chapter 8. At the beginning, the 5G network will have only one slice to support user connections, and gradually the number of slices will increase to specialize in the connection of objects, the connection of vehicles, the connection of machine tools, etc. Then, these slices could become enterprise networks, enabling the different sites of the same company to be interconnected.

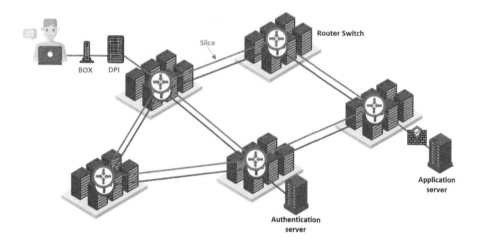

Figure 1.11. *Slicing. For a color version of this figure, see www.iste.co.uk/haddadou/edge.zip*

The step shown in Figure 1.12 is the complete virtualization of the physical infrastructure to allow the arrival of the digital infrastructure. All functions are virtualized, from the Internet box to the DPI, including the firewall and the authentication server, to name just a few.

12   Cloud and Edge Networking

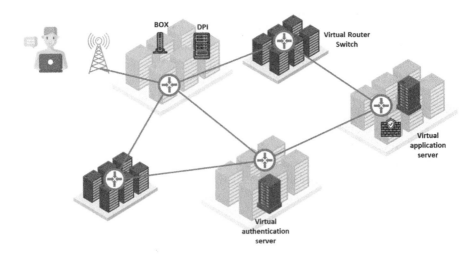

Figure 1.12. *Virtualization of all network functions. For a color version of this figure, see www.iste.co.uk/haddadou/edge.zip*

Let us look, for example, at the case of the Internet box that is being set up by the operators. The antenna cannot be virtualized and therefore it is necessary to keep a box that will contain the Wi-Fi antenna and/or the 5G antenna. Behind this antenna, an optical fiber connects the electromagnetic signal reception box to a data center of the operator. These centers are those being deployed for 5G. They will be located less than 10 km from the antenna to allow for an extremely short latency time, in the order of a millisecond. Indeed, a 20-km round trip at the speed of light on an optical fiber takes 0.1 ms. With the access time to the antenna and the processing time of a short request, we are in the order of a millisecond. This value corresponds to the maximum latency time for real-time applications such as the control of a vehicular network, between the start of braking of one vehicle and the start of braking of the next vehicle. Similarly, for the control of robots and machine tools, one millisecond is the recommended time. Similarly, for performing remote surgery where the surgeon needs to see what he is doing, a time lapse in the order of one millisecond is acceptable. The functions integrated in the Internet box are virtualized in the operator's data center. The name of the data centers of 5G operators is MEC (Multi-access Edge Computing), which succeeded the first definition of Mobile Edge Computing that referred only to mobile networks, while 5G is interested in all types of networks whether fixed or mobile. For example, a LAN (Local Area Network) using Wi-Fi is one of the connected systems in the 5G universe. The DPI function, which analyzes streams bit by bit, allowing the detection of anomalies by not recognizing the signatures of certain applications, is also virtualized in the Cloud. Similarly, the firewall and authentication server are virtualized in one of the operator's MEC data centers or possibly in a data center of a Cloud provider.

The question of where the virtual machines or containers are positioned must be asked. Today, we work with the four levels that are represented in Figure 1.13. The level called the Cloud represents the large data centers that are at the heart of the Internet, the core Clouds. The other three levels make up the Edge Cloud, which is abbreviated as the Edge. The Edge itself has three levels: the MEC, the Fog and the Embedded Edge.

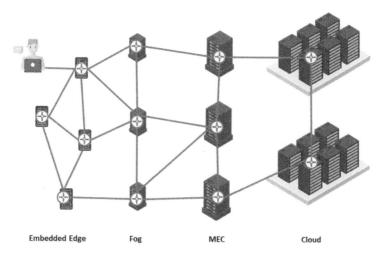

Figure 1.13. *The four levels of Cloud and Edge Networking. For a color version of this figure, see www.iste.co.uk/haddadou/edge.zip*

MEC is the furthest level from the customer, specified and realized in the 5G environment that introduces these data centers to define a digital infrastructure that will be built on top of the MEC data centers. 5G antennas or terrestrial access via Wi-Fi or other LAN techniques are connected over the MEC via fiber optics. The MEC data centers are intended to host all the virtual machines of the physical equipment and software located between the terminal equipment and the data center: signal processing, localization functions, control and management algorithms, environment autopilot, artificial intelligence processes, business applications, etc. In particular, there are real-time applications with strong constraints such as latency time, which must be in the millisecond range.

The name Fog Computing comes from the company CISCO, which in 2012 proposed to connect emerging objects (sensors, actuators and the like) on an intermediate data center to have a pre-processing before routing the selected information to a central Cloud to be processed. The term Fog has remained, and today it refers to data centers in companies but more particularly data centers on campuses, very close to the users, that is, a few hundred meters at most. The idea is to replace all physical equipment with virtual machines and to bring all business processes together as virtual machines or containers in the Fog data center. The latencies are very small, less than a millisecond.

The third level has several possible names including Embedded Edge, which we will use, as well as Skin, Mist or Far Edge. It is located very close to the user, within range of Wi-Fi or private 5G, which have the same range (a few tens of meters at most) since these two technologies transmit on free bands with the same constraints. The data centers are embedded computers of relatively low power at first. However, these embedded devices will become more and more powerful and will accept containerization and serverless technologies. The advantages of this solution are numerous: extremely low latency, data security while remaining in the user's lap and minimization of energy consumption. The objective is also to have a mobile environment, the embedded Edge can be in a vehicle, in a mobile object, in a smartphone or in a specific equipment in a user's pocket. The embedded Edge equipment is itself connected either to a more distant antenna such as an operator's 5G antenna or to other embedded Edges to form an embedded Cloud. This solution automates vehicular networks or oversees the control of mobile robots, and more globally may realize intelligent mobile environments.

1.4. Network Functions Virtualization

The basis of the technologies used in the previous sections comes from NFV (Network Functions Virtualization), which consists of virtualizing all the functions of the various boxes, such as a NAT device, a firewall, a DPI, a controller, a router, etc., as shown in Figure 1.14.

The problem with NFV comes from the potential non-compatibility of virtual machines between them. As a result, operators, who were at the origin of the request, wanted to standardize virtual machines to allow simple interconnection between operators. The first request for standardization was addressed to ETSI, the telecommunications standardization body for Europe, and at the request of many companies from all over the world, they opened the doors to a worldwide standardization. Furthermore, ETSI approached the Linux Foundation to create an open-source code reflecting this standardization of virtual machines. But a few months after the start, ETSI wished to go further by proposing an open-source software package associated with each function to allow them to interact to realize a complete and operational platform. This platform took the name of OPNFV (Open Platform Network Functions Virtualization) and, at the end of the development process on December 31, 2021, more than ten thousand people-years of work (i.e. the equivalent of ten thousand people working for 1 year) to realize this software. The final name that was given to it is Anuket, coming from the LF-Networking (Linux Foundation Networking). The project lasted 6 years with the development of intermediate releases from A to I.

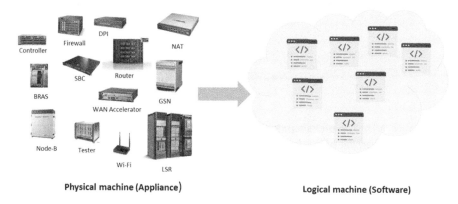

Figure 1.14. *Network Functions Virtualization (NFV). For a color version of this figure, see www.iste.co.uk/haddadou/edge.zip*

This Anuket platform of LF-Networking includes much open-source software of the Linux Foundation. The main work was to agglomerate all software while completing it. The general structure of this platform is described in Figure 1.15. It has three main parts:

– NFVI (NFV Infrastructure), which is the infrastructure element required to run virtual machines not only for networking but also for storage and computation.

– VNF (Virtual Network Function), which represents all available virtual functions from which the system will draw to perform the services requested by the users.

– NFV MANO (Management and Orchestration), which is the autopilot of the platform.

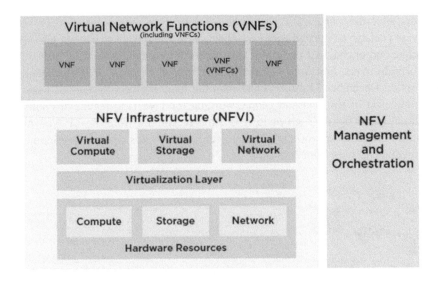

Figure 1.15. *The architecture of the LF-Networking Anuket platform. For a color version of this figure, see www.iste.co.uk/haddadou/edge.zip*

The more precise architecture of the LF-Networking Anuket platform will be described in a later chapter.

1.5. Conclusion

We have seen in this chapter an introduction to the digital infrastructure that has greatly simplified the network environments we knew with a strong centralization of functions, whether they are digital infrastructure functions like signal processing or routing or infrastructure service functions with control, management, intelligence, automation, etc. or application functions corresponding to the large applications requested by the users. This change is revolutionary and has led to Cloud Networking, which is slowly coming into place as many pieces of the puzzle are not yet really available, such as MEC data centers or slicing.

1.6. References

Antonopoulos, N. and Gilla, L. (2017). *Cloud Computing: Principles, Systems and Application*. Springer, New York.

Artasanchez, A. (2021). *AWS for Solutions Architects: Design Your Cloud Infrastructure by Implementing DevOps, Containers, and Amazon Web Services*. Packt Publishing, Birmingham.

Ben Jemaa, F., Pujolle, G., Pariente, M. (2016). Cloudlet- and NFV-based carrier Wi-Fi architecture for a wider range of services. *Annals of Telecommunications*, 71(11–12), 617–624.

Comer, D. (2021). *The Cloud Computing Book: The Future of Computing Explained*. CRC Press, Boca Raton.

Culkin, J. and Zazon, M. (2021). *AWS Cookbook: Recipes for Success on AWS*. O'Reilly, Sebastopol.

Dutt, D. (2019). *Cloud Native Data Center Networking: Architecture, Protocols, and Tools*. O'Reilly, Sebastopol.

Fox, R. and Hao, W. (2017). *Internet Infrastructure: Networking, Web Services, and Cloud Computing*. CRC Press, Boca Raton.

Gessert, F., Wingerath, W., Ritter, N. (2020). *Fast and Scalable Cloud Data Management*. Springer, Cham.

Gray, K. and Nadeau, T.D. (2016). *Network Function Virtualization*. Morgan Kaufmann, Burlington.

Halabi, S. (2019). *Hyperconverged Infrastructure Data Centers: Demystifying HCI*. Cisco Press, Indianapolis.

He, Y., Ren, J., Yu, G., Cai, Y. (2019). D2D communications meet mobile Edge computing for enhanced computation capacity in cellular networks. *IEEE Transactions on Wireless Communications*, 18(3), 1750–1763.

Kraemer, F.A., Braten, A.E., Tamkittikhun, N., Palma, N. (2017). Fog computing in healthcare – A review and discussion. *IEEE Access*, 5, 9206–9222.

Mach, P. and Becvar, Z. (2017). Mobile Edge computing: A survey on architecture and computation offloading. *IEEE Communications Surveys & Tutorials*, 19(3), 1628–1656.

Mao, Y., You, C., Zhang, J., Huang, K., Letaief, K.B. (2017). A survey on mobile Edge computing: The communication perspective. *IEEE Communications Surveys Tutorials*, 19(4), 2322–2358.

Moura, J. and Hutchison, D. (2019). Game theory for multi-access Edge computing: Survey, use cases, and future trends. *IEEE Communications Surveys Tutorials*, 21(1), 260–288.

Mouradian, C., Naboulsi, D., Yangui, C., Glitho, R.H., Morrow, M.J., Polakos, P.A. (2018). A comprehensive survey on fog computing: State-of-the-art and research challenges. *IEEE Communications Surveys Tutorials*, 20(1), 416–464.

Mukherjee, M., Shu, L., Wang, D. (2018). Survey of fog computing: Fundamental, network applications, and research challenges. *IEEE Communications Surveys Tutorials*, 20(3), 1826–1857.

Olaoye, A. (2022). *Beginning DevOps on AWS for iOS Development*. Apress/Springer Nature, Cham.

Perera, C., Qin, Y., Estrell, J.C.A., Reiff-Marganiec, S., Vasilakos, A.V. (2017). Fog computing for sustainable smart cities: A survey. *ACM Computing Surveys*, 50(3), 1–43.

Satyanarayanan, M. (2017). The emergence of Edge computing. *Computer*, 50(1), 30–39.

Shaukat, U., Ahmed, E., Anwar, Z., Xia, F. (2016). Cloudlet deployment in local wireless networks: Motivation, architectures, applications, and open challenges. *Journal of Network and Computer Applications*, 62, 18–40.

Sujata, D., Subhendu, K., Ajith, A., Yulan, L. (2021). *Advanced Soft Computing Techniques in Data Science, IoT and Cloud Computing*. Springer, Cham.

Vaquero, L.M. and Rodero-Merino, L. (2014). Finding your way in the fog: Towards a comprehensive definition of fog computing. *ACM SIGCOMM Computer Communication Review*, 44(5) 27–32.

Zburivsky, D. and Partnet, L. (2021). *Designing Cloud Data Platforms*. Manning Publications, Shelter Island.

2

The Cloud Continuum

2.1. Cloud Continuum levels

The Cloud Continuum refers to the continuity in the types of Clouds that can be set up. We defined four levels in the previous chapter, which we will detail a little more with respect to their main functionality: the ability to move a virtual machine from one level to another to position it in the place that optimizes a set of criteria ranging from performance to energy consumption through security, availability and other functions to be determined according to the optimization criteria requested by the customer.

These virtual machine migrations can be performed either hot or cold. In the first case, an identical machine is started on the server receiving the migration, then the configuration is transported from the sending virtual machine to the receiving virtual machine and finally the starting machine is shut down. The two virtual machines can run synchronously or quasi-synchronously for a short period of time. In the case of a cold migration, the starting virtual machine is stopped and it will restart only after the transfer of the virtual machine or its configuration only if an identical virtual machine is available at the receiver.

The Cloud Core can be divided into two types of data centers: hyperscalers and core data centers. Hyperscalers are intended for massive computing, whether it is Cloud Computing or intelligent processes from Big Data or machine learning. The hyperscale infrastructure must also enable very high levels of performance as well as fail-safe availability through redundancy, ensuring high fault tolerance.

Core data centers are large, more traditional data centers with several hundred to several hundred thousand servers. These data centers are the heart of the Internet, far from the customers, and through their power allow quick work on phenomenal amounts of data.

Cloud and Edge Networking,
by Kamel HADDADOU and Guy PUJOLLE. © ISTE Ltd 2023.

Figure 2.1 shows five core data centers, surrounded by a belt of Multi-access Edge Computing (MEC) data centers. These MECs, which are being deployed by 5G operators, are necessary to define the infrastructure of this new generation of networks capable of virtualizing all processes from the antenna to the user, passing through the intermediate boxes.

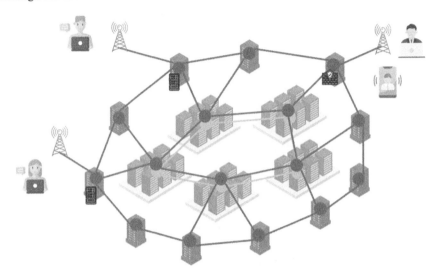

Figure 2.1. *The data center infrastructure of operators and Cloud providers. For a color version of this figure, see www.iste.co.uk/haddadou/edge.zip*

The goal of 5G operators is to support all the business data and processes that require very low latency. The GAFAM companies (Google, Amazon, Facebook, Apple and Microsoft), and more generally the big web companies, cannot handle these data because the latency times are too long compared to the needs. In reaction to this constraint, the big web companies are playing on several approaches to not stay out of these new markets, either by contracting with 5G operators to build the operators' MEC infrastructure or by going directly from the customer to their data centers using satellite constellations. The latter solution is however not satisfactory from a latency point of view, but it is a solution to maintain direct communication between the customer and the core data center.

The MEC data center gathers the connections to the different antennas that are within a radius of 10 km. It manages all the processes from signal processing to application services and control and management processes. This new paradigm has been developed to facilitate the digitalization of companies through this coupling with telecom operators.

Figure 2.2 allows us to get even closer to the terminal equipment by introducing the Fog Networking layer. As we saw in the previous chapter, this layer corresponds to the enterprise. It contains the data centers in which the company's virtual machines, such as switches or routers, security and control equipment, application services and artificial intelligence processes are virtualized.

The maximum distance between the endpoint and the Fog data center is the size of a corporate campus, which can be reduced to a single room. A local area network is required to carry the data from the end device to the data center. This network is very typically an Ethernet/Wi-Fi network, but networks using private 5G are rapidly emerging. Companies like Amazon Web Services (AWS) are marketing private 5G networks to get closer to customers. The case of AWS is very telling : the company is encouraging its customers to install small Fog data centers that are managed by Amazon and connected to the entire Amazon digital infrastructure.

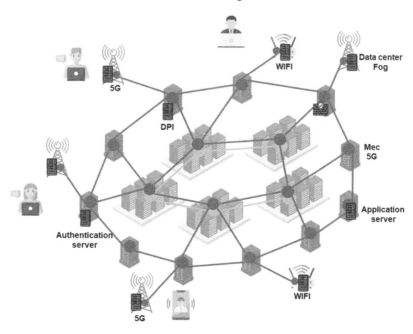

Figure 2.2. *The Fog layer in the digital infrastructure. For a color version of this figure, see www.iste.co.uk/haddadou/edge.zip*

Several types of local networks are possible. The most classical one comes from an Ethernet/Wi-Fi network connecting the client wirelessly on the Wi-Fi antenna and to go back to the data center through the Ethernet network. Another solution, as we have just seen, that should rise strongly comes from private 5G that uses private bands that might be available in some states, but more likely using free ISM (Industrial, Scientific and Medical) bands already used by Wi-Fi and other technologies like Bluetooth. The band that actually allows this introduction was opened in 2020 in the United States and late 2021 in Europe. It includes frequencies from 5.945 to 7.155 GHz and therefore covers 1.2 GHz. The maximum power being 200 mW, the range is quite similar to that of a 100 mW Wi-Fi but on a much lower band, that of 2.4 GHz. The range of private 5G should therefore be quite similar to that of Wi-Fi. Private 5G should be particularly useful for industrial manufacturing automation in competition with industrial networks. Its advantage comes from directional antennas enabling the simultaneous control of several machine tools or robots or cobots.

The last level, the Embedded Edge, puts the client within a short distance of the terminal equipment, that is, in the range of Wi-Fi. The embedded data center can embed an intelligent processor that accelerates artificial intelligence processes in the algorithmics of complex systems. It can also embed powerful processors capable of responding in real time with high reliability to support complex processes. An important example that can be cited for these embedded data centers comes from the control of vehicular networks. Behind this automated driving application lies all the data from the vehicular world. The volume of this data is growing exponentially and represents a fabulous potential windfall for those who will process it.

2.2. Cloud Continuum Networks

The Cloud Continuum consists of a suite of data centers, ranging from the infinitely small to the infinitely large, located anywhere from a few millimeters to thousands of kilometers from the user. On this platform, virtual machines, containers and functions must be urbanized, that is, precisely positioned on the different levels of the Cloud Continuum.

We have represented a Cloud Continuum in Figure 2.3 to reflect the different levels explained. The data center connections are generally made with the tier above or the tier below, but of course all cases are possible. For example, a Fog data center can very well be connected directly to a hyperscale data center.

Generally speaking, a latency of less than a tenth of a millisecond corresponds to an embedded data center, a millisecond or a little more to a MEC data center, and several tens of milliseconds to Cloud or hyperscale data centers.

A difficult problem, which has no simple solution, concerns urbanization as soon as several performance criteria come into play. Moreover, the optimization can change at any time and processes must be migrated from one level to another while taking into account the latency time to perform these migrations. Security and reliability criteria are particularly complex to take into account, but despite this they will be increasingly integrated into urbanization algorithms.

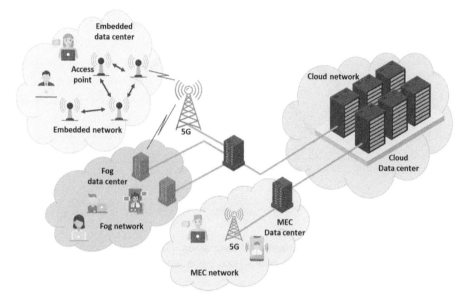

Figure 2.3. *A Cloud Continuum. For a color version of this figure, see www.iste.co.uk/haddadou/edge.zip*

2.3. The Cloud Continuum and the digitization of companies

The digitization of companies aims to produce new values in the world of production, in business models, in business software and more globally in the internal capabilities of companies to support new business processes. It is a method of value production based on the creation of consumer benefits through the use of digital technology.

Going digital is about rethinking the entire business model and understanding where the boundaries of value lie. Understanding these new boundaries may include creating entirely new businesses.

Figure 2.4 shows the different levels of the digital enterprise and their associated functions.

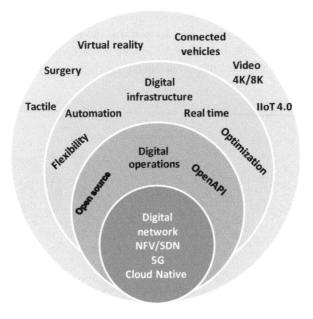

Figure 2.4. *The digitization of companies. For a color version of this figure, see www.iste.co.uk/haddadou/edge.zip*

The basis of the digital enterprise comes from the digital infrastructure that must be put in place to allow the introduction of the necessary intelligence to automate and optimize the functions that make the company live. This requires a basic physical environment that consists of antennas, optical fibers and data centers that form the Cloud Networking. The key words are 5G, NFV, SDN and Cloud Native Architecture.

This basic infrastructure enables digital operations in an open-source environment with open interfaces. All this leads to the digital infrastructure which must be automated and in real time. This infrastructure must support all business processes and services related to industry, audiovisual, games, metaverse, tactile applications, etc.

The importance of the digital infrastructure, shown in Figure 2.5, comes from the consolidation of all the functions needed by the business into the data center or data centers associated with the business. The other elements come from the connection of the equipment to the data centers.

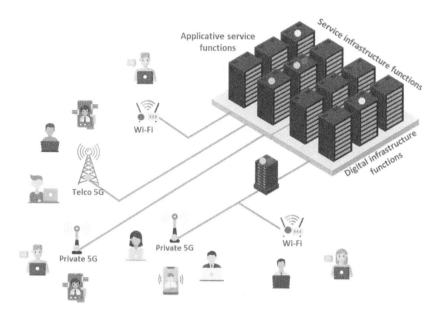

Figure 2.5. *The digital infrastructure. For a color version of this figure, see www.iste.co.uk/haddadou/edge.zip*

The advantage of this digital infrastructure is that it allows easy scalability of companies by adding virtual machines, at will, of any type and for any function.

2.4. Example of digital infrastructure

The big web companies were the first to set up a high-level digital infrastructure with a vision of the Cloud Continuum. They started with large Cloud or hyperscale data centers and then started to develop Edges especially in the enterprise context. Concerning MEC datacenters, big web companies cooperate with telcos. They are now tackling the embedded Edge, but there is still a lot to do in this direction, especially to decide how to get in. To illustrate this implementation, we will develop the case of Google but we would find very similar elements at Amazon and Microsoft.

Google deploys a world-wide network that is shown in Figure 2.6.

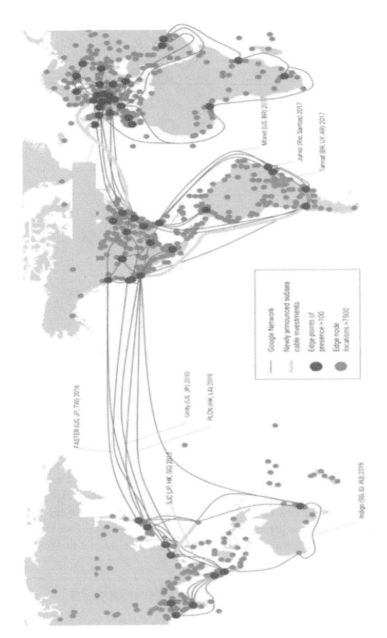

Figure 2.6. *The Google network (© Google). For a color version of this figure, see www.iste.co.uk/haddadou/edge.zip*

Figure 2.6 shows Google's points of presence corresponding to three levels of data centers: global data centers, regional data centers and area data centers. There are 27 very large data centers called Jupiter, the largest of which reaches one million servers and consumes 100 MW. We are in the 2 GW range of energy expenditure on the highest level network, the B4 network that connects the 27 Jupiter data centers. The total throughput is 100 Tbit/s, which represents 40 Tbit/s for each data center. There are 82 zones in total covering the world. Each zone has several data centers, which can be called Edge, to connect the customers. These data centers form Google's B2 network, which interconnects with the Internet and allows any user to reach Google's network.

Figure 2.7 represents this infrastructure.

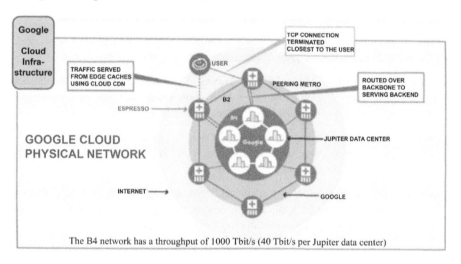

Figure 2.7. *Google's infrastructure. For a color version of this figure, see www.iste.co.uk/haddadou/edge.zip*

To manage this infrastructure, Google uses a technique called ZTN (Zero Touch Network). This architecture carries many features such as autoconfiguration, security and automation. The B4 network uses Google's proprietary SD-WAN technology with a triplicated control center.

2.5. Conclusion

The Cloud Continuum is a new digital generation for building intelligent and automatable environments. These environments must enable high flexibility and high reliability and provide security for the data used in them. This continuum is being put in place. However, there are still many issues to be resolved, such as strong security and optimal urbanization.

2.6. References

Abbas, N., Zhang, Y., Taherkordi, A., Skeie, T. (2018). Mobile edge computing: A survey. *IEEE Internet of Things Journal*, 5(1), 450–465.

Ahmed, E., Gani, A., Khan, M.K., Buyya, R., Khan, S.U. (2015a). Seamless application execution in mobile cloud computing: Motivation, taxonomy, and open challenges. *Journal of Network and Computer Applications*, 52, 154–172.

Ahmed, E., Gani, A., Sookhak, M., Ab Hamid, S.H., Xia, F. (2015b). Application optimization in mobile cloud computing: Motivation, taxonomies, and open challenges. *Journal of Network and Computer Applications*, 52, 52–68.

Ahmed, E., Ahmed, A., Yaqoob, I., Shuja, J., Gani, A., Imran, M., Shoaib, M. (2017). Bringing computation closer toward the user network: Is Edge computing the solution? *IEEE Communications Magazine*, 55(11), 138–144.

Ahmed, A., Naveed, A., Gani, A., Ab Hamid, S.H., Imran, M., Guizani, M. (2019). Process state synchronization-based application execution management for mobile edge/cloud computing. *Future Generation Computer Systems*, 91, 579–589.

Allam, H., Nassiri, N., Rajan, A., Ahmad, J. (2017). A critical overview of latest challenges and solutions of Mobile Cloud Computing. *Proceedings of the Second International Conference on Fog and Mobile Edge Computing (FMEC'17)*, Valencia, Spain.

Aslanpour, M.S., Gill, S.S., Adel, N.T. (2020). Performance evaluation metrics for cloud, fog and edge computing: A review, taxonomy, benchmarks and standards for future research. *Internet of Things*, 12, 100273.

Belabed, B., Secci, S., Pujolle, G., Medhi, D. (2015). Striking a balance between traffic engineering and virtual bridging in virtual machine placement. *IEEE Transactions on Network and Service Management*, 12(2), 202–216.

Bilal, K., Khalid, O., Erbad, A., Khan, S.U. (2018). Potentials, trends, and prospects in edge technologies: Fog, cloudlet, mobile edge, and micro data centers. *Computer Networks*, 130, 94–120.

Dinh, H.T., Lee, C., Niyato, D., Wang, P. (2013). A survey of mobile cloud computing: Architecture, applications, and approaches. *Wireless Communications and Mobile Computing*, 13, 1587–1611.

Drolia, U.R., Martins, U.R., Tan, J., Chheda, A., Sanghavi, M., Satyanarayanan, M., Bahl, P., Caceres, R., Davies, N. (2009). The case for VM-based cloudlets in mobile computing. *IEEE Pervasive Comput.*, 8(4), 14–23.

ETSI (2013). Network functions virtualisation (NFV); architectural framework v1.1. White Paper, ETSI, Sophia Antipolis, France.

ETSI (2014). Network functions virtualisation (NFV); architectural framework v1.2. White Paper, ETSI, Sophia Antipolis, France.

ETSI (2017). Network functions virtualisation (NFV) release 3: Evolution and ecosystem. Report on network slicing support with ETSI NFV architecture framework. White Paper, Sophia, Antipolis, France.

ETSI (2018a). MEC deployments in 4G and evolution towards 5G. White Paper, ETSI, Sophia Antipolis, France.

ETSI (2018b). MEC in an enterprise setting: A solution outline. White Paper, ETSI, Sophia Antipolis, France.

Frahim, J. and Josyula, V. (2017). *Intercloud: Solving Interoperability and Communication in a Cloud of Clouds*. Cisco Press, Indianapolis, USA.

Liu, H., Cao, L., Pei, T., Deng, Q., Zhu, J. (2020). A fast algorithm for energy-saving offloading with reliability and latency requirements in multi-access edge computing. *IEEE Access*, 8, 151–161.

Moraes, I.M., Mattos, D.M.F., Ferraz, L.H.G., Campista, M.E.M., Rubinstein, M.G., Costa, L.H.M.K., Amorim, M.D., Velloso, P.B., Duarte, O.C.M.B., Pujolle, G. (2014). FITS: A flexible virtual network testbed architecture. *Computer Networks*, 63(4), 221–237.

Moura, J. and Hutchison, D. (2019). Game theory for multi-access edge computing: Survey, use cases, and future trends. *IEEE Communications Surveys Tutorials*, 21(1), 260–288.

Pan, J. and McElhannon, J. (2018). Future Edge Cloud and Edge Computing for Internet of Things applications. *IEEE Internet of Things Journal*, 5(1), 439–449.

Takeda, A., Kimura, T., Hirata, K. (2019). Evaluation of Edge Cloud server placement for Edge computing environments. In *IEEE International Conference on Consumer Electronics – Taiwan (ICCETW)*.

Vaquero, L.M. and Rodero-Merino, M.L. (2014). Finding your way in the Fog: Towards a comprehensive definition of Fog computing. *ACM SIGCOMM Computer Communication Review*, 44, 27–32.

Yi, S., Li, C., Li, Q. (2015). A survey of Fog computing: Concepts, applications and issues. Workshop on mobile Big Data. In *Proceedings of the 2015 Workshop on Mobile Big Data*.

3
Digital Infrastructure Architecture

3.1. The evolution of enterprise information system architectures

There are major steps forward every 20 years or so in information systems architectures for businesses. In Figure 3.1, we have represented the major steps since 1960, and the digital infrastructure architecture corresponds to the years 2020–2040.

In the 1960s, there was a hardware infrastructure composed of computers ranging from mainframes to minicomputers. On these devices, application services written in languages adapted to the category of applications were running (and sometimes still are). From 1980 onward, infrastructure services appeared between the hardware infrastructure and the application services, that is, all the common processes used to control and manage the environment. From the 2000s onward, the first datacenters appeared, supporting a first generation of digital infrastructure and in particular storage virtualization. Infrastructure services continue to develop by providing more and more common functions to the enterprise such as security or resilience. Finally, the 2020s only consolidate this architecture by integrating Cloud Continuum into the standard infrastructure and developing many functions to completely virtualize all physical devices. Infrastructure services continue to increase with intelligence, automation and flexibility features. This increase is such that many functions that were individualized in application services are being integrated into infrastructure services. This reduces the complexity of the application services, which will need only to use the numerous functions of the infrastructure service.

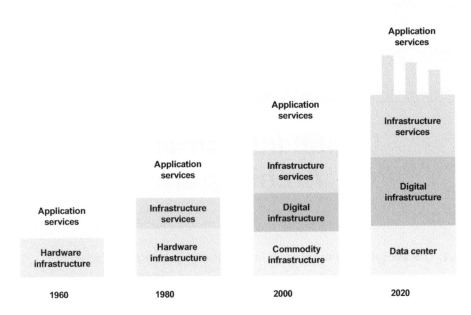

Figure 3.1. *Enterprise information systems architecture. For a color version of this figure, see www.iste.co.uk/haddadou/edge.zip*

In the following, we will detail the functions of the different levels of the architecture for the years 2020–2040.

Figure 3.2 details the layer that contains the application services, that is, more precisely the virtual machines, containers or functions for the application services. Classically, we find all the software that appears on our smartphone or personal computer. Many new applications will appear, not yet monopolized by the major web manufacturers. These new forms of software will come mainly from the 5G specifications. They are grouped into three major classes: applications requiring very high throughput, applications based on large numbers of objects and applications that require very small latency times (of the order of a millisecond) with high reliability. Within these three classes, we can cite Industry 4.0, which must automate industrial production, the smart grid, which automates electricity management and control processes, the Internet of Things and its extension called the Internet of Behaviors, vehicular networks with controls that require millisecond latencies, tactile networks with real-time management and haptic communications.

Figure 3.3 focuses on the second layer, which corresponds to infrastructure services. The applications at this level offer services that are common to the other layers. In particular, we find security, control, urbanization of services and microservices, resource allocation, intelligence through machine learning or Big Data analysis, digital twin services or management services. This layer tends to grow strongly by including more and more functions that are found today in the application layer. Instead of finding similar processes in each application service, these are pooled by a single application in the infrastructure services layer. This is the origin of the new vision of programming by functions instead of coding microservices. The number of functions available in this layer increases to simplify the development of new application services.

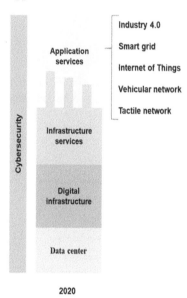

Figure 3.2. *Application services. For a color version of this figure, see www.iste.co.uk/haddadou/edge.zip*

Figure 3.4 shows the virtual machines of the digital infrastructure. The network machines mainly include routers, switches, firewalls, authentication servers, DPI (Deep Packet Inspection) but also the signal processing needed in Wi-Fi boxes or 4G or 5G antennas. It also includes location management and handover processing systems. For the 5G world, this layer deals with the controls and management of the systems such as the setting up of a D2D connection or the management of a private 5G antenna. We can place the virtualizations of the eNodeB electronic cabinets or the box of a Wi-Fi router here.

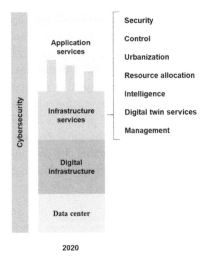

Figure 3.3. *Infrastructure services. For a color version of this figure, see www.iste.co.uk/haddadou/edge.zip*

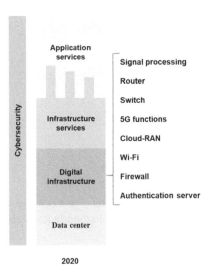

Figure 3.4. *The digital infrastructure. For a color version of this figure, see www.iste.co.uk/haddadou/edge.zip*

For all virtual machines, security must be provided to avoid cyber-attacks. Let us review some new directions in this area. First of all, a strong axis concerns the SIM cards which will be replaced by iSIM (integrated SIM) cards. These new SIM cards are very small, less than 1 mm², and they are integrated in the SoC (System as a Chip) of machines that have a processor, whatever the size of the processor. This solution will make it possible to secure objects that are easily attackable today.

These iSIMs can be eSIMs (embedded SIMs), that is, they can have multiple profiles that are all, with the exception of the main profile, virtualized in secure equipment located in companies that may be far away. In other words, we are dealing with a remote card that is accessed via a secure link from the smartphone or tablet.

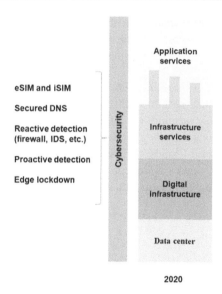

Figure 3.5. *Cybersecurity and independence. For a color version of this figure, see www.iste.co.uk/haddadou/edge.zip*

Another direction comes from secure DNS resolvers. Indeed, today the main DNS resolvers used by companies are located at the big web and telecom operators who can easily spy on the companies' requests. It is necessary to secure DNS resolvers by using for example DoH (DNS over HTTPS) or DNS over TLS.

On the attack side, in contrast to the classic solution of reactive detection of an attack, carried out by a firewall or an IDS, we must oppose proactive solutions that monitor the Internet to detect future attacks in advance, for example, by monitoring domain names.

Another solution that is on the rise concerns the containment of the Edge by cutting off the Internet connection that brings the attacks. This requires repatriating all the data and processes needed for businesses to function properly to the Edge.

For example, attacks against hospitals could have been avoided, like ransomware, by disconnecting from the Internet and working in a partitioned way. This solution is obviously not easy to implement, if not impossible for some configurations that need to reach data all over the world. However, statistical studies show that two-thirds of companies could very well work locally.

3.2. The Open Infrastructure Foundation architecture

There are two main infrastructure models that have similarities and elements in common, the Open Infrastructure Foundation (OIF) and the Cloud Native. We will take a look at them to see the characteristics.

Let us start with OIF. This architecture is based on OpenStack which is an open-source Cloud management system. In 2019, the developers of OpenStack realized that there was a whole section of infrastructure missing in addition to the management system, in particular the Edge part, containers that were replacing hypervisor systems, hybrid Clouds integrating a public and a private part, artificial intelligence processes and agile programming systems. These different characteristics can be found in Figure 3.6, which form the key elements of OIF.

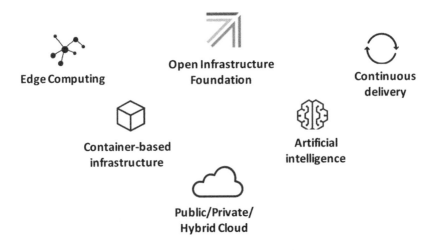

Figure 3.6. *The basic elements of OIF architecture. For a color version of this figure, see www.iste.co.uk/haddadou/edge.zip*

Before we look at the main additions to OpenStack to form this new foundation, let us explain the structure of this management system, which is the main management system adopted around the world. It is mostly used by Cloud providers and telecom operators. Enterprises often prefer commercial management software such as VMware.

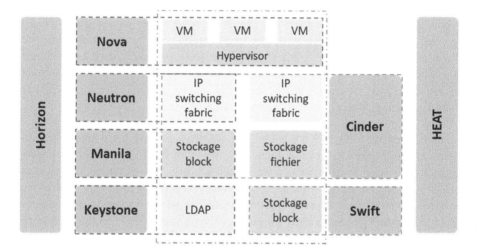

Figure 3.7. *An overview of OpenStack modules. For a color version of this figure, see www.iste.co.uk/haddadou/edge.zip*

OpenStack is a Cloud operating system that controls computing, storage and network resources in a data center, all managed and provisioned via application programming interfaces (APIs) with common authentication mechanisms. A dashboard is available, giving administrators control while allowing their users to provision resources via a web interface. Beyond the standard Infrastructure as a Service (IaaS) functionality, additional components provide orchestration, fault management and service management. The software guarantees high availability of user applications.

Most proprietary solutions are run on software designed by and for a single organization. OpenStack is different. This management software is operated across multiple Cloud providers. OpenStack can be managed as a public or private Cloud, operated by a single organization with its own goals and making its own decisions. These Cloud management systems may have overlapping sets of users. This leads to some specific requirements for OpenStack.

Development done on a vendor's Cloud operating under OpenStack must be transportable to another OpenStack system. In other words, an application must be able to be migrated with only minor modifications to various public and private OpenStack Clouds. Where the requirements for deployment differ (e.g. between a public and private Cloud), OpenStack should strive to design mechanisms that can be used in either environment so that applications can be migrated from one OpenStack Cloud to another. Deployment implementation details, such as the choice of different backend or configuration drivers that may be specific, should not intrude on the consumer experience to the extent possible. In particular, these choices should not change the behavior of the application interfaces.

Improvements to OpenStack Cloud management software are compatible with older versions, making it easy to upgrade to newer versions. However, when users interact with several different Clouds simultaneously, this assumption no longer holds. A user interacting with a newer version of OpenStack may then have problems running their applications on an older version of OpenStack. Services developed under OpenStack must therefore evolve in such a way that they can work properly for both older and newer clients.

Not all OpenStack Clouds include the same set of services, and deploying new services requires additional work on the part of Cloud operators. While OpenStack encourages new services under development to reuse functionality from other already supported services, reuse should not be maximized by adding hardware dependencies. Choosing to add a hardware dependency always involves a trade-off between design simplicity and operational flexibility. Projects should add a hardware dependency when they judge that it ultimately benefits users, for example, by reducing the area of security-sensitive code, reducing the possibility of duplicate bugs, enabling desirable properties such as scalability or resiliency, or increasing the team's development speed. Particular weight should be given to the security benefits for operators and users.

A region in an OpenStack Cloud is defined as a distinct set of service endpoints in the Keystone service catalog, allowing a registered user to access any region in the Cloud from the same authentication URL. In contrast, resource groupings defined by hardware or physical data center topology are under the control of individual Cloud operators. For example, many Clouds include the concept of "availability zones", or groupings within a region that share no common point of failure. OpenStack software has no way to enforce this property for multiple Clouds. OpenStack projects are encouraged to allow operators to create arbitrary, hierarchical groupings of the resources they manage and to avoid assigning physical meanings to groupings.

The following design goals represent the functionality that OpenStack services as a whole provide to applications and users. Each service or feature is not expected to address goals already listed. Instead, any service that contributes to one or more of the goals is likely to help advance the mission of the OpenStack project.

OpenStack does not assume the existence of an operational data center; it provides the tools to operate a data center and make its resources available to consumers. There is no layer required under OpenStack as a whole. It provides the abstractions needed to manage external systems such as compute, storage and networking hardware, domain name system and identity management systems. The OpenStack APIs provide a consistent interface to these systems, each of which can potentially be implemented by a variety of vendors and open-source projects. This broad base provides an abstraction that can host more specialized services, both from OpenStack itself and from third parties.

OpenStack supports and encourages additional abstraction layers including platforms as a service, serverless compute platforms and container orchestration engines. OpenStack provides tools to support tight integration into an OpenStack Cloud of popular third-party open-source projects that provide these layers.

OpenStack projects that include an abstraction layer on multiple potential backend services can also offer this abstraction layer as a standalone entity, to be consumed by external services independently of a true OpenStack Cloud.

For any service from specialized hardware, OpenStack aims to provide a vendor-independent API that gives consumers software-based control over resource allocation in a multi-tenant environment. This is not limited to virtual servers, but can include things such as storage, routers, load balancers, firewalls, HSMs, GPGPUs, FPGAs or ASICs (e.g. video codecs). Some of these categories of hardware may have purely software equivalents that can be used behind the same API, allowing applications to be portable even to Clouds that do not have specialized hardware in these cases.

OpenStack strives to provide application developers with interfaces that allow them, in principle, to efficiently scale from very small to very large workloads without re-architecting their applications. This allows consumers to use capacity as needed and share the underlying resources with other applications and tenants, rather than allocating discrete blocks to particular applications and wasting excess capacity in the blocks.

In an environment with unreliable hardware, making an application reliable is difficult and, for small applications, very expensive. OpenStack aims to provide primitives that allow developers to create reliable applications on top. The underlying resources can be shared between applications and *tenants* so that the cost is amortized between them, rather than requiring each application to pay the full cost. The existence of these primitives allows some other services to be simpler and more scalable.

OpenStack does not impose any particular deployment model or architecture on applications. Each application has unique requirements, and OpenStack accommodates them by allowing services to be linked together in the user space via public APIs rather than through hardwired actions taken in the background that only support predefined deployment models.

This allows application developers to customize any application using client-side glue. OpenStack services are integrated enough that they can be connected together by the Cloud client without requiring client-side interaction beyond the initial wiring.

Security models must enable both types of interaction, between OpenStack services and between applications and OpenStack services, in both directions. They must also allow the Cloud consumer to delegate only the minimum privileges necessary to allow the application to function as intended, and allow for regular revocation and replacement of credentials to maintain the highest possible security in an environment where Internet-connected machines are susceptible to compromise at any time.

Certain components of an application (for example, databases) often benefit from a specialized skill set to operate them. By abstracting the management of some of these more common components behind an API, OpenStack allows the relationship between these components and the rest of the application to be formalized. For organizations that have access to specialists, this allows them to cover more applications by working centrally. For other organizations, it allows them to access specialized skills they might not otherwise have access to, via a public or managed OpenStack Cloud.

Not all reusable components of an application guarantee their own OpenStack service. Suitable candidates typically have complex configuration, ongoing lifecycle management needs, and sophisticated OpenStack infrastructure requirements (such as virtual server cluster management).

A graphical interface is often the best way for new users to approach a Cloud and for users in general to experience unfamiliar areas of it. The graphical presentation of options and workflows makes it much easier to discover functionality than reading the API or command-line interface (CLI) documentation. A graphical interface is also often the best way, even for power users and Cloud operators, to get a broad overview of the state of their Cloud resources and to visualize the relationships between them. For these reasons, in addition to the API and any other user interface, OpenStack should include a web-based GUI (Graphical User Interface).

To OpenStack we need to add the Edge infrastructure to achieve a complete environment. Among the additional open-source software to OpenStack in OIF is Magma, which is open-source software for the Edge infrastructure offering both cellular (5G) and Wi-Fi services. There is also OpenInfra Labs, which brings together open-source software. For example, there is the Wenjua project, which provides services to shorten the introduction time, often important for deploying artificial intelligence. There is also the well-known Kubernetes container orchestration software that actually comes from the Cloud Native, which we will look at as the main architecture for the Edge and Cloud. This orchestrator is described in Figure 3.8.

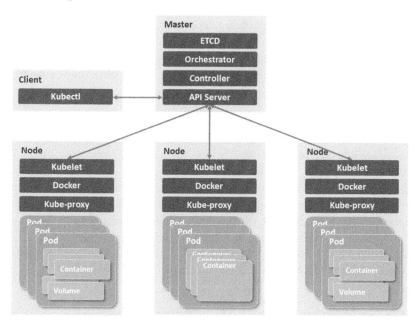

Figure 3.8. *The Kubernetes orchestrator architecture. For a color version of this figure, see www.iste.co.uk/haddadou/edge.zip*

The basic element is the pod. The pod is used to host an application instance. A pod is a Kubernetes abstraction that represents a group of one or more application containers (such as Docker) and a number of resources associated with those containers. These resources include storage, an IP address and information about the execution of each container.

A pod models an application-specific "logical host" that can contain different tightly coupled application containers. For example, a pod may include both a container for the application as well as a container for feeding data to be published by the web server. Containers in a pod share an IP address and a port number. They are always collocated and coplanar and run in a shared context on the same node.

Pods are the atomic unit on the Kubernetes platform. When deploying an application on Kubernetes, that deployment creates pods with containers inside (as opposed to creating containers directly). Each pod is bound to the node where it is scheduled and remains there until termination or deletion. If a node fails, identical pods are scheduled on other available nodes in the cluster.

A pod always runs on a node. A node is a working machine in Kubernetes and can be a virtual or physical machine, depending on the cluster. Each node is managed by the control plane. A node can have multiple pods, and the Kubernetes control plane automatically manages the scheduling of pods on the nodes in the cluster. The automatic scheduling of the control plane takes into account the resources available on each node.

Each Kubernetes node runs Kubelet, a process responsible for communication between the Kubernetes control plane and the node. The node manages the pods and containers running on a machine.

The ETCD module is an open-source key store used to store and manage critical information that distributed systems need to keep running. Specifically, this module manages configuration data, state data and metadata for Kubernetes.

The master contains the previous modules but also a scheduler and a controller that form the control system of Kubernetes.

3.3. The Cloud Native Computing Foundation architecture

The Cloud Native Computing Foundation (CNCF) is a project of the Linux Foundation that was founded in 2015 to help advance container technology and push this technology into the enterprise. This project was announced at the same time as Kubernetes 1.0, which, as we just saw, is an open-source container cluster manager that was provided to the Linux Foundation by Google as a starter technology.

Figure 3.9 introduces the four basic elements of the Cloud Native. First, there are containers and microservices, which we have already explained in the previous paragraphs. To these must be added the DevOps technologies of agile programming that enable the continuous delivery of programs and their updates.

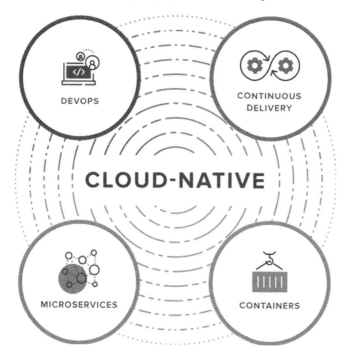

Figure 3.9. *The four basic elements of Cloud Native. For a color version of this figure, see www.iste.co.uk/haddadou/edge.zip*

A DevOps approach is one of many techniques used by IT staff to execute IT projects that meet business needs. DevOps can coexist with Agile software development. Some IT professionals believe that simply combining Dev and Ops is not enough, and the term DevOps should explicitly include business (BizDevOps), security (DevSecOps) or other areas.

DevOps is a philosophy that fosters better communication and deeper collaboration between development and operations teams in an organization. In its narrowest interpretation, DevOps describes the adoption of iterative software development, automation, deployment and maintenance of programmable infrastructure. The term also covers culture changes, such as building trust and cohesion between developers and system administrators and aligning technology projects with business needs. DevOps can change the software delivery chain, services, jobs and IT tools.

Although DevOps is not a technology per se, DevOps environments generally apply common methodologies. These include tools for integration, continuous integration or continuous deployment (CI/CD), with an emphasis on task automation. Also included are systems and tools that support DevOps adoption, including real-time monitoring, incident management, configuration management and collaboration platforms.

Figure 3.10 introduces the complete architecture of the Cloud Native. There are four parts. The bottom part deals with the infrastructure, which can be public, private or hybrid. The middle part supports containerization or function-based techniques, with all the service, management and application layers above, that is, function integration and application interfaces for application management. On the right are all the entities that can be observed and finally, on the left of the figure are all the management and control processes.

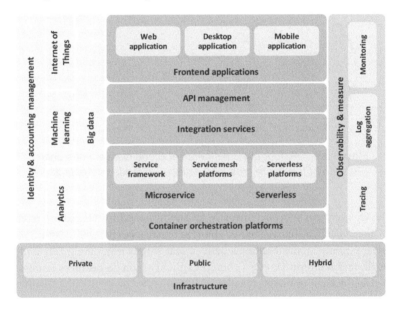

Figure 3.10. *Cloud Native Architecture. For a color version of this figure, see www.iste.co.uk/haddadou/edge.zip*

There are other infrastructure projects for both the Edge and the Cloud. We can cite the work of the ONF (Open Network Foundation), which has defined open architectures around Cloud networking. Examples include the SEBA (SDN-Enabled Broadband Access), Treillis (Open-source multi-purpose leaf-spine fabric), ODTN (Open and Disaggregated Transport Network), OMEC (ONF's Open-Source Mobile Evolved Packet Core) and AETHER (Open-source platform for 5G, LTE and Edge as a Cloud service).

Another development that could also play a role, at least in Europe, comes from the EECC (European Edge Computing Consortium) project, which defines a rather complex three-dimensional architecture. This architecture aims to pool research and development efforts to provide technology stacks for Edge data centers. In particular, the components must be homogeneous to enable simple interconnection of Edge data centers in industrial IoT and smart manufacturing.

Finally, we can mention an architecture coming from the operators: the Telco Edge Cloud. The Edge inherits most of the principles, mechanisms and tools of the traditional Cloud, but provides customers with additional benefits, as the Edge brings proximity to users' devices, enabling features such as low latency, which is relevant for reaction time sensitive applications. The Operator Platform Group (OPG) has defined Edge capabilities for operators, resulting in the Telco Edge Cloud (TEC) Taskforce.

The initiative aims to create an interoperable architecture of the Edges of many telecommunications operators. In addition, this architecture should facilitate the work of developers of applications located on the Edge. To support this initiative, the GSMA Platforms Group has developed the required specifications.

The two major architectures, OIF and Cloud Native, are based on containers and microservices. As shown in Figure 3.11, this architecture succeeded virtual machines, which in turn followed Bare Metal servers. The future, from the time of writing, will be found in function-based techniques.

Containers, microservices and Kubernetes have now become the standard for Cloud computing. What remains to be defined are the necessary software layers between the developer and the Kubernetes orchestrator. The solution that is developing and that should become the standard is serverless: the developer does not have to worry about the machine or the containers on which the programs will run. The two major solutions of today and tomorrow are CaaS (Container as a Service) and FaaS (Function as a Service). To distinguish between them and place them in relation to the major IaaS, PaaS and SaaS standards, we need to examine the exact architecture, which is described in Figure 3.12.

Trend towards Serverless

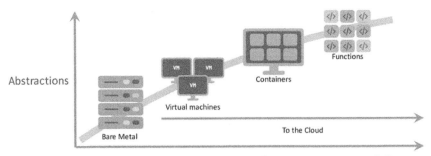

Figure 3.11. *Cloud Computing Architectures. For a color version of this figure, see www.iste.co.uk/haddadou/edge.zip*

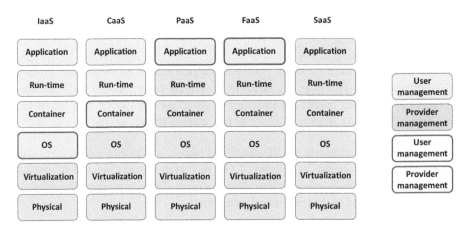

Figure 3.12. *CaaS and FaaS architecture. For a color version of this figure, see www.iste.co.uk/haddadou/edge.zip*

We can see that CaaS is an architecture in which the container comes from the Cloud provider but the management of the container is handled by the user. In FaaS,

Digital Infrastructure Architecture   47

all layers come from the provider except for the application, which comes from the provider's functions but is managed by the user.

All architectures integrate containers and Kubertenes (or the equivalent since there are other possibilities, though not widely used). The serverless technology is therefore placed on top of Kubertenes, but the latter becomes transparent to the user. There are many implementations of the serverless mode. For example, on top of Kubernetes, we can put the open-source software Knative (Cloud-native serverless) which allows the deployment of serverless applications and their execution on all Kubernetes platforms. Knative provides a set of middleware components to create modern, container-based applications that can run anywhere: on-premises, in the Cloud or even in a third-party data center.

Knative can be replaced by OpenFaaS which works from functions in a serverless mode supported by Kubernetes and Docker.

Apache OpenWhisk is a third solution for implementing a serverless environment. It is an open-source distributed platform that runs functions in response to events at any scale. OpenWhisk manages infrastructure, servers and scaling using Docker containers to focus all developer efforts on building applications in a very simple way. The OpenWhisk platform supports a programming model in which developers write functional logic, called Actions, in any supported programming language that can be dynamically scheduled and executed in response to events from external sources or HTTP requests. The project includes a REST API-based CLI as well as other tools to support packaging, catalog services, and numerous container deployment options.

Finally, the three major Cloud Computing companies of the GAFAM, Amazon, Google and Microsoft, have their own software stack to achieve serverless. For example, Amazon uses AWS Lambda. AWS Lambda is a serverless computing service that allows you to run code without worrying about the servers on which the code will run. Applications run under AWS Lambda without any administration. The term "lambda" refers to serverless functions. Most programming languages can be used.

Figure 3.13 summarizes the three architectures that follow one another but are all in use today: the monolithic remains very present, the microservice is at the top, and the functions are arriving quickly.

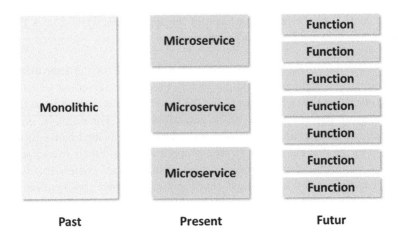

Figure 3.13. *The three architectures for Cloud Computing. For a color version of this figure, see www.iste.co.uk/haddadou/edge.zip*

Another representation of these different solutions is described in Figure 3.14 with the main differences between these architectures.

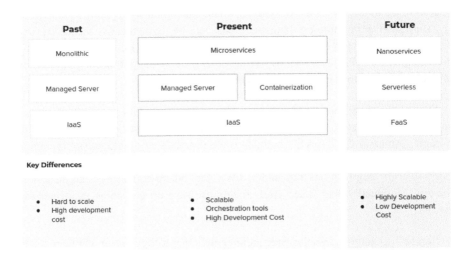

Figure 3.14. *Comparison of the three Cloud Computing architectures. For a color version of this figure, see www.iste.co.uk/haddadou/edge.zip*

3.4. Gaia-X

To conclude this chapter, we will introduce the European Gaia-X project, which is an attempt to react against GAFAM companies to try to free up data and allow them to simply migrate from one provider's data center to another provider's data center. Figure 3.15 introduces the main elements of this project. On the right side of the figure are the providers and on the left side are the consumers. The basic idea is to be able to migrate virtual machines, containers or functions from one data center to another for various reasons ranging from cost to performance, security, latency, etc. In this scheme, we find the urbanization of virtual machines that must be positioned in the best possible place with respect to a whole set of constraints. The migration can go from a core network to an Edge and vice versa. One of the important paradigms is Cloud federation, that is, a set of Clouds whose providers have agreed to carry out these migrations in a flexible and secure way.

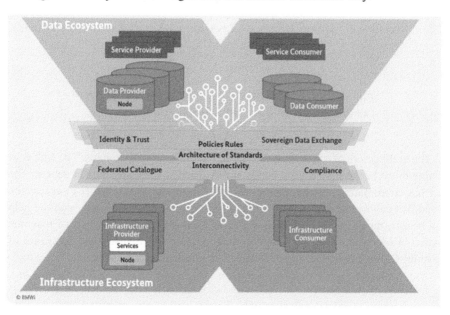

Figure 3.15. *The architecture of the Gaia-X project (© Gaïa-X). For a color version of this figure, see www.iste.co.uk/haddadou/edge.zip*

A Cloud Federation is an interconnection of Clouds with the objective of load balancing and handling traffic peaks but also to have access to specific services available on only certain Clouds. The Cloud Federation also allows a provider to use the points of presence of another provider and to be able to expand abroad, for example, by using this property.

Figure 3.16 shows schematically a federation of Clouds that are directly interconnected, but it is possible to have star architectures around a controller that automatically directs flows to the right Cloud.

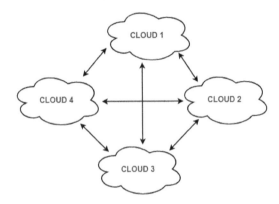

Figure 3.16. *A federation of Clouds. For a color version of this figure, see www.iste.co.uk/haddadou/edge.zip*

The architecture of a federation of Clouds has three basic elements. The first is to be able to exchange information between the Clouds and/or the user. In general, the user has coordinating software that allows a customer of the company to access the Cloud that can handle his request with the right quality of service and, with the help of a broker, the cheapest cost. This software itself happens to be a virtual machine located in one of the Clouds. This coordinating software is inside the controller of the multi-Cloud environment. This software can also act as a broker.

The broker coordinates the exchange between users and the available services provided by the Cloud Controller. The broker knows the current costs, demand patterns and available Cloud providers. This information is periodically updated by the Cloud Coordinator.

The second element of the Cloud federation is the controller of the Clouds that form the multi-Cloud. This controller allocates resources from the different Clouds to remote users based on the quality of service required by the users and the credits that these users have for accessing the different Clouds. The Cloud providers and their access are managed by the controller.

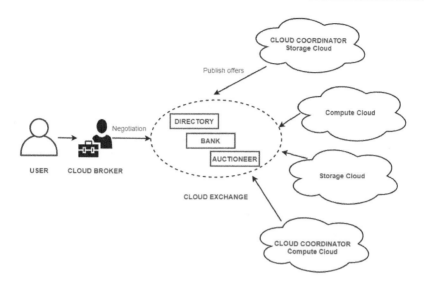

Figure 3.17. *Architecture of a federation of Clouds. For a color version of this figure, see www.iste.co.uk/haddadou/edge.zip*

The third element comes from the Cloud broker who interacts with the controller, analyzes the fit with the service levels and resources offered by the different Cloud providers. The Cloud broker finalizes the most appropriate agreement for the customer.

The most important properties of the federation of Clouds are the following.

– Users can interact with the architecture in a centralized or distributed manner. In centralized interaction, the user interacts with a broker to mediate between them and the organization. Distributed interaction allows the user to interact directly with the Clouds in the federation.

– The Cloud federation can be used in different scenarios, such as commercial or non-commercial use.

– The visibility of a federation of Clouds allows the user to know the organization of each Cloud in the federated environment.

– Cloud federation can be monitored in two different ways. The first is the MaaS (Monitoring as a Service) solution that provides information to help monitor the services contracted to the user. The second way is to choose a global monitoring solution that allows the monitoring and management of the federation of Clouds.

– Suppliers participating in the federation publish their offers to a central entity. The user interacts with this central entity to check prices and respond to one of the offers.

– Marketing objects such as infrastructure, software and platform must pass through the controller when supported by the Cloud federation.

Cloud federation has many advantages that can be summarized in the following properties.

– It minimizes energy consumption.

– It increases the availability of information or calculations to be performed.

– It minimizes the time and cost of suppliers: this is due to dynamic scalability.

– It connects various Cloud service providers on a global scale. Providers can buy and sell services on demand.

– It provides a scaling of resources.

However, there are still many challenges that need to be addressed to enable simple deployment of Cloud federations. The main challenges are the following:

– In a Cloud federation, it is common to have more than one inbound request processing provider. In such cases, there must be a system in place to distribute inbound requests evenly among the Cloud service providers.

– The increasing demands of a Cloud federation are leading to a more heterogeneous infrastructure, with interoperability being a concern. This concern often leads to a challenge for Cloud users to select relevant Cloud service providers. In general, the choice made binds users to a particular Cloud service provider.

– Cloud federation means the need to build a seamless Cloud environment that can interact with people, devices, application interfaces and many other entities.

Several technologies have been developed to facilitate Cloud federations. The main ones that have been implemented are OpenNebula, Aneka, Eucalyptus and Gaia-X, which we have already briefly examined. OpenNebula is a management platform for heterogeneous distributed data center infrastructures. This system manages Cloud resources by leveraging technology assets to protect transactions and data. An API has also been developed to access the management system. The second technology is Aneka, which provides services and components that optimize interactions with the Cloud. The third technology comes from Eucalyptus, which defines the pooling of compute, storage and network resources that can be scaled up or down according to the application's workload. Eucalyptus is an open-source software that manages the storage, network and compute of a Cloud environment.

In the fourth solution, instead of building a new Cloud, Gaia-X offers a network infrastructure capable of connecting multiple heterogeneous Cloud systems and all different types of data ecosystems, while complying with government requirements and European Union laws.

In summary, the emerging infrastructure is therefore a networked system, connecting multiple Cloud service providers and data sources from private companies, universities, research centers and NGOs, while ensuring openness, transparency and trust.

Gaia-X is an open-source environment, under Apache license (version 2.0). It will be maintained in the GitHub repository. It is mainly written in Python and Java. In addition, rust, C++, Go and Node.js can be used if needed. Since the architecture needs to be encapsulated in microservices, a container-based solution such as Docker, is used with Kubernetes as the container orchestrator.

Policies define the set of rules for Gaia-X. Assets and resources describe the objects and properties of the Gaia-X ecosystem and together make up a service offering. An asset can be a data asset, a software asset, a node or an interconnection asset. A set of policies is linked to each asset. An asset does not have an interface to the outside world. In contrast, resources have an external interface to a function or service. They are also linked to certain policies. The difference between resources and assets is that resources represent the elements that are necessary for the provided assets. Resources can be explained as internal service instances not available for commands. For example, the operating instance providing a data set is a resource.

The three roles that a participant in the Gaia-X ecosystem plays are the provider, the consumer and the federator.

While a consumer is a participant who can search for service offerings and consume service instances to enable digital offerings for end users, a provider is a participant who offers a service instance. A service instance includes a self-description and technical policies that use different resources and have different assets. The federator has a rather special role. It is responsible for the services of the federation, which are independent of each other. A federation designates a set of actors interacting directly or indirectly, producing or providing assets and related resources.

The federation services are the core services of Gaia-X. They should enable the entire system to function perfectly. They are designed as microservices that provide the essential functionality needed to manage the entire infrastructure.

Federation services include four groups of services needed to enable the federation of assets, resources, participants and interactions between ecosystems. The four service groups are identity and trust, federated catalog, sovereign data exchange and compliance. In addition, the infrastructure will be available through a portal.

3.5. Conclusion

Digital infrastructure architectures form the basis of any system. The struggle to impose a standard is quite unequal between the different entities impacted by these architectures. The GAFAM is largely at the head of any proposal given their weight representing 70% of commercialized Cloud services. If we add a few more American companies such as IBM and Salesforce, as well as the two or three major Chinese web companies, we reach 95%. The rest of the world is therefore excessively underpowered to change the course of the proposals. The Cloud Native seems to be well on its way to continue being the great standard of the domain.

3.6. References

Ahmed, E., Akhunzada, A., Whaiduzzaman, M., Gani, A., Ab Hamid, S.H., Buyya, R. (2015). Network-centric performance analysis of runtime application migration in mobile Cloud computing. *Simulation Modeling Practice and Theory*, 50, 42–56.

Ahmed, E., Naveed, A., Gani A., Ab Hamid, S.H., Imran, M., Guizani, M. (2017). Process state synchronization for mobility support in mobile Cloud computing. *International Conference on Communications (ICC)*, 38, 1–6.

Al-Roomi, M., Al-Ebrahim, S., Buqrais, S., Ahmad, I. (2013). Cloud computing pricing models: A survey. *International Journal of Grid and Distributed Computing*, 6(5), 93–106.

Bahman, J., Jingtao, S., Rajiv, R. (2020). Serverless architecture for Edge computing. *Edge Computing: Models, Technologies and Applications*. In *Institution of Engineering and Technology*, Western Sydney University Research Direct.

Baldini, I., Castro, P., Chang, K., Cheng, P., Fink, S., Ishakian, V., Mitchell, N., Muthusamy, V., Rabbah, R., Slominski, A., Suter, P. (2017). Serverless computing: Current trends and open problems. In *Research Advances in Cloud Computing*, Chaudhary, S., Somani, G., Buyya, R. (eds). Springer, Singapore.

Bao, W., Yuan, D., Yang, Z., Wang, S., Li, W., Zhou, B.B., Zomaya, A.Y. (2017). Follow me fog: Toward seamless handover timing schemes in a fog computing environment. *IEEE Communications Magazine*, 55(11), 72–78.

Bourguiba, M., Haddadou, K., El Korbi, I., Pujolle, G. (2014). Improving network I/O virtualization for Cloud computing. *IEEE Trans. Parallel Distrib. Syst.*, 25(3), 673–681.

Cicconetti, C., Conti, M., Passarella, A., Sabella, D. (2020). Toward distributed computing environments with serverless solutions in Edge systems. *IEEE Communications Magazine*, 58(3), 40–46.

Djennane, N., Aoudjit, R., Bouzefrane, S. (2018). Energy-efficient algorithm for load balancing and VMS reassignment in data centers. In *6th International Conference on Future Internet of Things and Cloud Workshops, IEEE Proceedings*.

Eismann, S., Scheuner, J., van Eyk, E., Schwinger, M., Grohmann, J., Herbst, N., Abad, C., Losup, A. (2020). Serverless applications: Why, when, and how? *IEEE Software*, 38, 32–39.

Fajjari, I., Aitsaadi, N., Pióro, M., Pujolle, G. (2014). A new virtual network static embedding strategy within the Cloud's private backbone network. *Computer Networks*, 62(7), 69–88.

Fox, R. and Hao, W. (2017). *Internet Infrastructure: Networking, Web Services, and Cloud Computing*. CRC Press, Boca Raton.

Geng, H. (2014). *Data Center Handbook*. John Wiley & Sons, New York.

Ghobaei-Arani, M., Souri, A., Rahmanian, A.A. (2020). Resource management approaches in fog computing: A comprehensive review. *Journal of Grid Computing*, 18(1), 1–42.

Glikson, A., Nastic, S., Dustdar, S. (2017). Deviceless Edge computing: Extending serverless computing to the Edge of the network. *Proceedings of the 10th ACM International Systems and Storage Conference*, IBM Research, New York.

Hadji, M., Djenane, N., Aoudjit, R., Bouzefrane, S. (2016). A new scalable and energy efficient algorithm for VMs reassignment in Cloud data centers. In *IEEE International Conference on Future Internet of Things and Cloud Workshops (FiCloudW)*, Vienna.

Han, B., Gopalakrishnan, V., Ji, L., Lee, S. (2015). Network function virtualization: Challenges and opportunities for innovations. *IEEE Communications Magazine*, 53(2), 90–97.

Hassan, N.S., Gillani, S., Ahmed, E., Yaqoob, I., Imran, M. (2018). The role of Edge computing in internet of things. *IEEE Communications Magazine*, 56(11), 110–115.

Laghrissi, A. and Taleb, T. (2018). A survey on the placement of virtual resources and virtual network functions. *IEEE Communications Surveys & Tutorials*, 21(2), 1409–1434.

Leivadeas, A., Kesidis, G., Ibnkahla, M., Lambadaris, I. (2019). VNF placement optimization at the Edge and Cloud. *Future Internet*, 11(3), 69.

Li, C., Xue, Y., Wang, J., Zhang, W., Li, T. (2018). Edge-oriented computing paradigms: A survey on architecture design and system management. *ACM Computing Surveys (CSUR)*, 51(2), 1–34.

Lin, L., Li, P., Liao, X., Jin, H., Zhang, Y. (2019). Echo: An Edge-centric code offloading system with quality of service guarantee. *IEEE Access*, 7(37), 5905–5917.

Mechtri, M., Ghribi, C., Zeghlache, D. (2016). A scalable algorithm for the placement of service function chains. *IEEE Transactions on Network and Service Management*, 13(3), 533–546.

Rhoton, J., De Clercq, J., Novak, F. (2014). *OpenStack Cloud Computing: Architecture Guide*. RP Publisher, USA and UK.

Saboowala, H., Abid, M., Modali, S. (2013). *Designing Networks and Services for the Cloud: Delivering Business-grade Cloud Applications and Services*. Cisco Press, Indianapolis.

Satyanarayanan, M., Bahl, P., Caceres, R., Davies, N. (2009). The case for VM-based cloudlets in mobile computing. *IEEE Pervasive Computing*, 8(4), 14–23.

Shiraz, M., Gani, A., Khokhar, R.H., Buyya R. (2013). A review on distributed application processing frameworks in smart mobile devices for mobile Cloud computing. *IEEE Commun. Surv. Tutor.*, 15(3), 1294–1313.

Slim, F., Guillemin, F., Gravey, A., Hadjadj-Aoul, Y. (2017). Towards a dynamic adaptive placement of virtual network functions under ONAP. In *IEEE Conference on Network Function Virtualization and Software Defined Networks (NFV-SDN)*, Berlin.

Tavakoli-Someh, S. and Rezvani, M.H. (2019) Utilization-aware virtual network function placement using nsga-ii evolutionary computing. *IEEE Conference on Knowledge Based Engineering and Innovation*, Tehran.

Wang, P., Yao, C., Zheng, Z., Sun, G., Song, L. (2017). Joint task assignment, transmission and computing resource allocation in multi-layer mobile Edge computing systems. *IEEE Internet of Things Journal*, 32, 2872–2884.

Wang, T., Zhang, G., Liu, A., Bhuiyan, M.Z.A., Jin, Q. (2019). A secure IoT service architecture with an efficient balance dynamics based on Cloud and Edge computing. *IEEE Internet of Things Journal*, 6(3), 4831–4843.

Woldeyohannes, T., Mohammadkhan, A., Ramakrishnan, K., Jiang, Y. (2018). ClusPR: Balancing multiple objectives at scale for NFV resource allocation. *IEEE Trans. on Net. and Serv. Manag.*, 15(4), 1307–1321.

4

Open-Source Architectures for Edge and Cloud Networking

4.1. Organizations and the main open sources

Among the organizations developing open-source software, the most important for networks and the Cloud is the Linux Foundation, which is in charge of gathering development forces to work on Cloud, containers and virtualization, the Internet of Things and embedded and orchestration, but also on platforms and applications. The industry is firstly interested not in open-source software but in open-source hardware that is, the Open Compute Project (OCP), which we will examine in more detail in the following lines. Then, among the important concerns of the manufacturers, we find the following open source: Anuket/OPNFV/CNTT/LF Networking, open-source Serverless, Magma, EdgeX Foundry, ONAP, OpenDaylight, O-RAN, Open vSwitch, PNDA and Smart Edge Open. This list is not exhaustive but these are some of the most requested software, and we will describe them in a little more detail.

4.2. The main open-source projects

The objective of the OCP project is to provide open access to specifications of hardware and more precisely of hardware intended to build data centers of various sizes. By making these specifications free, it allows specialized companies, the ODM (Original Design Manufacturers), to produce these machines in white label or white box. This project was initially launched by Facebook with the idea of making data centers at a lower cost. Moreover, the equipment has been designed to consume as little energy as possible or to recover heat. There are all kinds of equipment, from very low-cost cards to more specific equipment for data center sizes ranging from very small to very large. The success of OCP is obvious, and more and more companies are choosing this solution and building low-cost data centers. However, it should be noted that these sales are made without maintenance contracts, which can greatly increase the cost of using a third-party company for maintenance.

Anuket is a Linux Foundation project that is the largest part of the LF-Networking environment. This project is the culmination of one of the largest projects ever launched: OPNFV (Open Platform NFV). The goal is to create an open-source, verified platform using virtualization. OPNFV is an architecture for a carrier-grade platform. Carrier-grade indicates among other things a high availability to reach an availability of at least 99.999%, that is to say, a few minutes of downtime per year corresponding to interruptions and maintenance. This platform has undergone some change to achieve finalization. Indeed, several versions were proposed, in particular one for telecommunications operators: the CNTT platform (Cloud iNfrastructure Telco Task Force). Fortunately, a convergence took place to gather the different versions under the name of Anuket, which is the main part of the Linux Foundation Networking. In addition to Anuket, LF-Networking is in charge of two other major projects, Cloud Native and ONAP (Open Network Automation Platform), which we will see a little later.

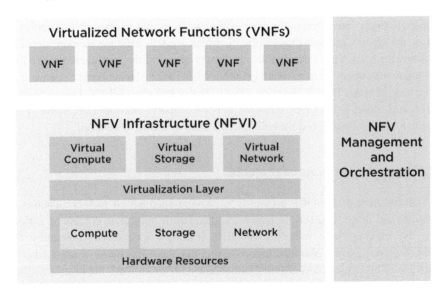

Figure 4.1. *The Anuket/OPNFV platform (OPNFV). For a color version of this figure, see www.iste.co.uk/haddadou/edge.zip*

The initial phase of OPNFV was to define the NFV Infrastructure and Virtualized Infrastructure Management (VIM) of the environment. The APIs for NFV were also to be standardized. Overall, the first phase was to end with the definition of the infrastructure based on VNF (Virtualized Network Functions) and MANO (Management and Network Orchestration). After this initial phase of a few months, the European standards body ETSI, which was managing the project, decided to broaden the scope of the platform by proposing open-source software for everything needed to have a complete and operational platform. The schematic architecture of OPNFV is described in Figure 4.2.

This platform is made up of three main parts: MANO, which handles management and orchestration; NFVI, which defines the virtualization-based infrastructure and finally the virtual machines that are used, the VNFs (Virtual Network Function).

A slightly more detailed version is shown in Figure 4.2.

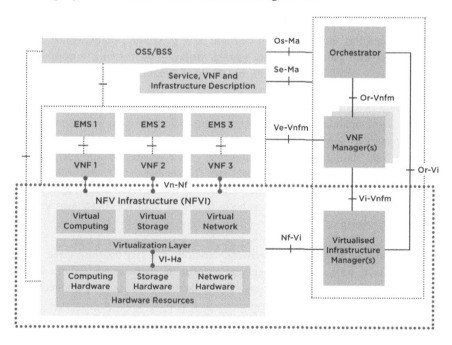

Figure 4.2. *The full version of the Anuket/OPNFV platform (OPNFV). For a color version of this figure, see www.iste.co.uk/haddadou/edge.zip*

The NFV orchestrator (NFVO) aims to realize the automatic setting up of networks. The orchestrator must be able to choose the virtual machines able to perform the requested service. To do this, the virtual machines must be chained, that is, put one behind the other to form a chain representing the order in which they will be crossed by the application's packet flow. Finally, the virtual machines must be urbanized, that is, placed on the physical machines to provide the service.

The NFV Manager takes care of the management of the VNF instances, that is, the virtual machines, their configuration and the reporting of the events coming from the virtual machines and their local management system.

Finally, the VIM (Virtualized Infrastructure Manager) takes care of the control and management of the NFV infrastructure (NFVI). To do this, it controls and manages the network, computing and storage resources. Finally, it collects and sends performance measurements to the VNFM.

ONAP (Open Network Automation Protocol) is one of the most important projects for the future because it concerns the automation of networks. This automatic control is the basis for the whole new generation of networks. In the mid-2010s, several parallel projects started working on network autopilot: the Open-O projects, from developments by China Telecom, Huawei and ZTE, and OpenECOMP, the open-source version of ECOMP and the SDN platform from AT&T. These two projects came together in 2017 to form ONAP. The focus was on network design, creation, orchestration, automation and lifecycle management. The main companies leading this project were AT&T, Amdocs, Orange, Huawei, ZTE, VMware, China Mobile, Bell Canada and Intel. The releases followed each other at a rate of two per year. The first, Amsterdam (November 2017), was followed by Beijing, Casablanca, Dublin, El Alto, Frankfurt, Guilin, Honolulu, Istanbul and Jakarta. The move to Cloud Native to be in line with Anuket/OPNFV was enacted from the Guilin version.

ONAP aims to automate the network environment. To do this, three major subsystems have been grouped together to form the network autopilot: service creation, orchestration, and operations. These three subsystems are connected as shown in Figure 4.3.

Figure 4.4 shows a more detailed version of the ONAP architecture. Each subsystem has many modules. The creation of services is done with the main module SDC (Service Design and Creation) which manages the element resources (infrastructure, network and application), services taking into account all the information available to start, destroy, update and manage the services, users' products taking into account the costs and finally the product groupings, taking into account the specifications and the recommended configurations.

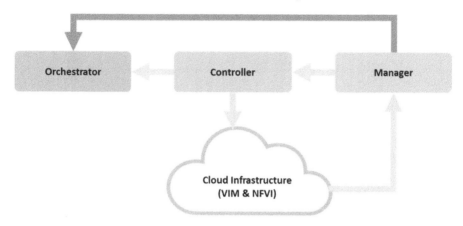

Figure 4.3. *The three ONAP subsystems. For a color version of this figure, see www.iste.co.uk/haddadou/edge.zip*

The orchestration subsystem contains the service orchestrator and the various ONAP autopilot controllers such as the SDN controller (SDN-C) whose choice is OpenDaylight, the application controller (APPC), the virtual function controller (VF-C) and the infrastructure controller (Infra-C) which can support multiple infrastructures.

The third subsystem concerns the operations of the network. It contains the DCAE (Data Collection Analytics & Events) module with the Holmes sub-module for alarm correlation and the NFV-O sub-module for monitoring.

The modules MSB (MicroService Bus) and DMAAP (Data Movement as a Platform) are complementary. MSB provides the application interface for registration, discovery, routing and gateway access to microservices. DMAAP, on the other hand, handles messages and file transfers between microservices.

Finally, the CLAMP (Closed Loop Automation Management Platform) module manages the feedback loops to update the configurations in case of a change in the network state.

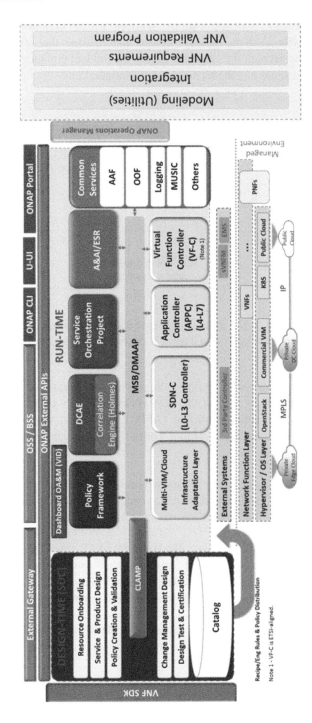

Figure 4.4. *The ONAP architecture (ONAP). For a color version of this figure, see www.iste.co.uk/haddadou/edge.zip*

The policy module takes care of the rule sets to implement, distribute and maintain them. These policy rules include virtual machine urbanization, data management, access control, actions to be executed, interactions, etc.

To conclude this section on ONAP, it is worth mentioning that ONAP is the centerpiece of the new generation of networks whose objective is to automate the operation of the network, storage and computing, from the creation of the service to its deployment and operation.

Open vSwitch is by far the most widely used virtual switch in the virtual network world. Almost all devices use this open-source software because of its many features. It can run as a switch with a hypervisor or as a distributed switch on multiple physical servers. It is the default switch in XenServer 6.0, Xen Cloud Platform and supports KVM. Open vSwitch is integrated into many Cloud systems such as OpenStack, OpenQRM, OpenNebula and oVirt. It is distributed by Ubuntu, Debian, Fedora Linux, FreeBSD, etc. A view of the Open vSwitch architecture is given in Figure 4.5.

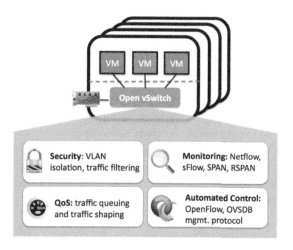

Figure 4.5. *The Open vSwitch virtual switch (OvS). For a color version of this figure, see www.iste.co.uk/haddadou/edge.zip*

Open vSwitch is used in several products and runs in many large production environments. The Open vSwitch community also manages the Open Virtual Network (OVN) project. OVN complements existing OVS features by adding native support for virtual network abstractions, such as overlay layers, Layer 2 and Layer 3 security virtual machines, overlay tunnels and gateways to the virtualized world.

The OpenDaylight (ODL) platform is a project, hosted by the Linux Foundation, to build an open-source platform to realize a controller to configure SDN switches through a southbound interface. An example of the architecture of the OpenDaylight platform is shown in Figure 4.6. This architecture corresponds to the Oxygen and Fluorine releases in 2018, Neon and Sodium in 2019, Magnesium and Aluminum in 2020, Silicon and Phosphorus in 2021 and Sulfur and Chlorine in 2022.

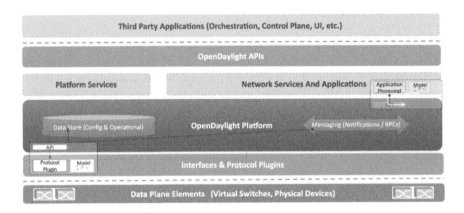

Figure 4.6. *Architecture of an OpenDaylight platform (OpenDaylight). For a color version of this figure, see www.iste.co.uk/haddadou/edge.zip*

Because the OpenDaylight platform is multiprotocol and modular, users can create a custom SDN controller to meet their specific needs. This modular, multiprotocol approach provides the ability to choose a single southbound interface or select multiple protocols to solve complex problems. The platform supports all southbound interfaces found in the literature both commercially and in open source, including OpenFlow, OVSDB, NetConf and BGP.

OpenDaylight also provides all of the northbound interfaces used by applications. These applications use the controller to collect information about the network and perform analysis, and then use it to create new rules to apply to the network.

The OpenDaylight controller gives rise to a Java virtual machine. This means that it can be deployed on just about any hardware and on any platform that supports Java. The controller uses the following tools:

– Maven, for managing the automation of the production of Java software projects;

– OSGi, for dynamic loading and packaging of JAR files;

– a JAVA interface, for listening to events, specifications and training models;

– REST API, to implement the topology manager, user tracking, flow implementation and associated static routing.

The Oxygen release includes a plugin for the P4 southbound interface to support data plane abstraction in the OpenDaylight platform. The P4 southbound interface signaling protocol, which is sometimes seen as the successor to OpenFlow, has been added to the signaling system already available in OpenDaylight. In addition to the P4 plug-in, the Oxygen version is dedicated to container orchestration. It includes a Kubernetes plug-in for orchestrating container-based workloads and extensions to the Neutron northbound interface for environments with multiple containers to host virtual machines.

OpenDaylight provides a BGP stack with enhancements in support for BGP/CEP and BGP/MPLS multicast, making OpenDaylight a mature solution for realizing an SD-WAN controller.

Several features have been added through OpenDaylight releases to improve support for network virtualization in Cloud, Fog and Edge embedded environments. This includes IPv6 support, stateful and stateless security group setup, and SR-IOV (Single Root I/O Virtualization) hardware offload for OvS (Open vSwitch) that greatly improves virtual switch performance. This work was developed largely for the OpenStack environment and is now being used to integrate OpenDaylight with the Kubernetes container orchestration engine.

Service Function Chaining (SFC) updates to the *Fluorine* version of OpenDaylight accelerate service provisioning by automatically chaining virtual machines.

To make OpenDaylight operational on large networks, development of eastbound–westbound interfaces is required. This is one of the features developed in the mid-2019 Neon release. With this in mind, OpenDaylight offers an inter SDN Controller Communication (SDNi) interface that uses the BGP4 protocol. This SDNi interface has become the standard for eastbound–westbound interfaces.

In conclusion, OpenDaylight is strengthening its commitment to other related open-source projects such as OpenStack, Kubernetes (k8s), Anuket/OPNFV and ONAP. In addition, the Neon release and subsequent releases provide a solid foundation for the new ONAP SDN Controller for Radio (SDN-R) project targeting the configuration of 5G environments. The OpenDaylight community has also begun an early collaboration with the Network Service Mesh (NSM) project that uses a new approach to solving complex L2/L3 use cases in Kubernetes.

An increasingly used accelerator comes from the FD.io (Fast Data input/output) project. It gave rise to open-source software to achieve very high performance in input/output of networking equipment. This project was launched in 2016. The Vector Packet Processing (VPP) core module provides code that runs in user space on x86, ARM and Power architectures. VPP's design is hardware, kernel and deployment independent (bare metal, hypervisor and container). VPP pushes the limits in terms of performance. Independent tests show that FD.io software, by using VPP, can improve performance by a factor of at least two during normal operation and even more when the system is congested. A diagram of the FD-IO operation is shown in Figure 4.7.

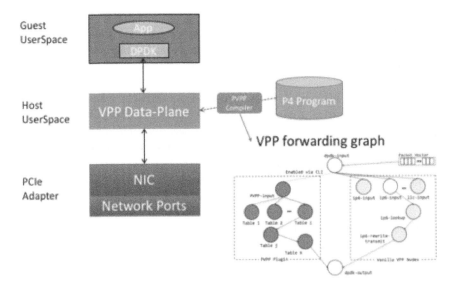

Figure 4.7. *FD.io accelerator architecture (FD.io). For a color version of this figure, see www.iste.co.uk/haddadou/edge.zip*

VPP processes the flow of a single session with the largest number of available packets from the network IO layer and simultaneously executes all these packets waiting for processing. Since all these packets have the same supervisory information, it is very simple to process them together. So, instead of processing the first packet on the entire input/output graph and then moving on to the second packet once the processing of the first packet has been completed, VPP processes all packets of the same session that are located in a node of the graph. This allows the system to process all the packets of the same vector much faster since the supervision area gives information in the first packet that is the same as in the following packets.

PNDA (Platform for Network Data Analytics) is a platform for performing big data analytics on network information using multiple open-source software. PNDA offers an innovative approach to collecting, processing and analyzing big data. The platform features a simplified data pipeline that makes it easy to get the right data at the right time. By decoupling data sources from data users, it is possible to integrate data sources and make them available to any application that wants to process them. Such platforms can perform data processing based on the arriving specific streams. On the other hand, client applications can use one of the interfaces to issue structured queries or consume streams directly.

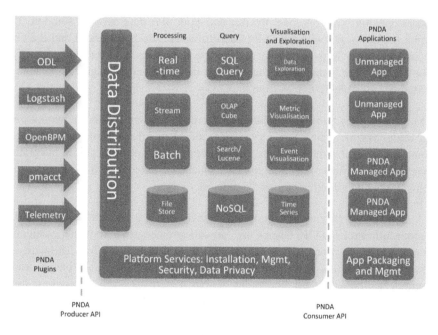

Figure 4.8. *The PNDA platform (PNDA). For a color version of this figure, see www.iste.co.uk/haddadou/edge.zip*

The PNDA platform is inspired by Big Data processing architecture models. It stores data in the rawest form possible in a resilient, distributed file system. PNDA provides the tools to process data continuously in real time and to perform deep analysis on large datasets. The architecture of PNDA is shown in Figure 4.8.

There are many plugins in PNDA. The most important ones are shown in Figure 4.8. There are the flows coming from the OpenDaylight controller, which itself retrieves all the network statistics through the southbound interface and in particular using OpenFlow. The second possibility is Logstash which is a tool for collecting, analyzing and storing logs. It is usually associated with ElasticSearch, a distributed search engine. There is also OpenBPM (Business Process Management) which encapsulates BGP messages from one or more BGP peers in a single TCP stream to one or more databases. The pmacct plugin is a set of monitoring tools that, among other things, export network flows. More generally, all telemetry software can be used to introduce data into the PNDA platform.

The Magma project aims to build an open-source software platform to provide network operators with an open, flexible and extensible core network solution. The Magma project works with the OIF (Open Infrastructure Foundation) and the OAI (OpenAirInterface Software Alliance). These cooperations have led to the founding of the Magma Core Foundation. The source code for one of the main components of the Magma core (MME) comes from the OAI.

Magma enables operators to build and expand modern mobile networks on a large scale. Magma features an access-independent mobile network core, advanced automation, control and network management tools. Magma also enables integration with existing LTE networks through virtual machines, but also with Wi-Fi networks. Magma is also interested in private 5G networks.

A number of Magma community members are also collaborating with the Telecom Infra Project (TIP) Open Core Network project group to define, create, test and deploy core networks that leverage Magma software as well as hardware and software solutions disaggregated by the Open TIP. Magma also has a strong presence in automating all operations within a network such as configuration, software updates and device provisioning. Magma enables better connectivity by allowing operators to extend their capacity and range by using LTE, 5G, Wi-Fi and CBRS. Magma also offers a cellular network service with an open source network in an equipment manufacturer offering proprietary solutions. Magma software also enables operators to manage their networks more efficiently with more automation, less downtime, better predictability and more agility to add new services.

EdgeX Foundry is an open-source project that facilitates interoperability between objects and applications that reside in the Edge. EdgeX Foundry provides all the protocols needed to achieve control, management and security of the platform. The platform is open and vendor-neutral and accelerates application development by providing modular reference services for analyzing and sharing data from objects. These properties make it very easy to develop new services based on data from the Internet of Things. Moreover, it is easy to add artificial intelligence and automation processes.

Akraino mirrors LF-Networking by representing LF-Edge (Linux Foundation Edge). It is almost as important a piece as LF-Networking. Akraino is a set of open infrastructure and applications for the Edge, covering a wide variety of use cases, including 5G, AI, Edge, IaaS/PaaS, IoT, that fit both Edge Computing vendors and enterprises that want to grow their Edge architecture. This environment was created by the Akraino community and focuses exclusively on the Edge. Akraino focuses on availability, capacity, security and continuity. Akraino is suitable for different types of Edge Computing architecture whether it is virtual machines, containers, microservices or functions.

Intel® Smart Edge Open is the successor to Open Network Edge Services Software (OpenNESS) and is a software package under the umbrella of the Linux Foundation. This project focuses on the development of a software toolkit for the creation of platforms for Edge computing. The objective is to accelerate the development of Edge solutions by hosting network functions alongside AI, media processing functions, resource optimization and security functions. This toolkit is based on the Cloud Native Platform. There is software to integrate 5G, to support accelerators like DPDK, to perform performance measurement and associated monitoring, etc.

4.3. Conclusion

Open source has become prevalent in the world of networking and more specifically Cloud Networking. We have introduced some of the software that is becoming the standard used in Edge and Cloud networking. It should be noted, however, that commercial products could come back in force. Indeed, open-source software has been developed for the most part by network and telecom equipment manufacturers, to define interfaces on which they will be able to build commercial software that perfectly respect these interfaces. These forms of commercial software could be more powerful, more secure, more available, more automated, etc.

4.4. References

Abbasi, A.A. and Jin, H. (2018). v-Mapper: An application-aware resource consolidation scheme for cloud data centers. *Future Internet*, 10(9), 90.

Barlow, M. (2017). *Data Structures and Transmission: Research, Technology and Applications*. Nova, New York.

Gkioulos, V., Gunleifsen, H., Weldehawaryat, G.K. (2018). A systematic literature review on military software defined networks. *Future Internet*, 10(9), 88.

Glikson, A., Nastic, S., Dustdar, S. (2017). Deviceless edge computing: Extending serverless computing to the edge of the network. *Proceedings of the 10th ACM International Systems and Storage Conference*, Haifa.

Gunleifsen, H., Gkioulos, V., Kemmerich, T. (2018). A tiered control plane model for service function chaining isolation. *Future Internet*, 10(6), 46.

Hakak, S., Noor, N.F.M., Ayub, M.N., Affal, H., Hussin, N., Imran, M., Ejaz, A. (2019). Cloud-assisted gamification for education and learning recent advances and challenges. *Computers & Electrical Engineering*, 74, 22–34.

Hall, A. and Ramachandran, U. (2019). An execution model for serverless functions at the edge. In *Proceedings of the International Conference on Internet of Things Design and Implementation*, Montreal.

Jiang, J.W., Lan, T., Ha, S., Chen, M., Chiang, M. (2012). Joint VM placement and routing for data center traffic engineering. In *Proceedings IEEE INFOCOM*, Orlando, FL.

Kirci, P. (2017). *Ubiquitous and Cloud Computing: Ubiquitous Computing, Resource Management and Efficiency in Cloud Computing Environments*. IGI Global, Hershey, PA.

Leivadeas, A., Kesidis, G., Ibnkahla, M., Lambadaris, I. (2019). VNF placement optimization at the edge and cloud. *Future Internet*, 11(3), 69.

Li, J., Altman, E., Touati, C. (2015). A general SDN-based IoT framework with NVF implementation. *ZTE Communications*, 13(3), 42–45.

Massonet, P., Deru, L., Achour, A., Dupont, S., Croisez, L.M., Levin, A., Villari, M. (2017). Security in lightweight network function virtualisation for federated cloud and IoT. In *IEEE 5th International Conference on Future Internet of Things and Cloud (FiCloud)*, Prague.

Medhat, A.M., Taleb, T., Elmangoush, A., Carella, G.A., Covaci, S., Magedanz, T. (2016). Service function chaining in next generation networks: State of the art and research challenges. *IEEE Communications Magazine*, 55(2), 216–223.

Open Edge Computing Initiative (2019). Open Edge Computing [Online]. Available at: https://www.openedgecomputing.org/.

Ramo, V., Lavacca, F.G., Catena, T., Polverini, M., Cianfrani, A. (2019). Effectiveness of segment routing technology in reducing the bandwidth and cloud resources provisioning times in network function virtualization architectures. *Future Internet*, 11(3), 71.

Rebello, G.A.F., Alvarenga, I.D., Sanz, I.J., Duarte, O.C.M. (2019). BSec NFVO: A blockchain based security for network function virtualization orchestration. In *IEEE International Conference on Communications (ICC)*.

Yao, J., Han, Z., Sohail, M., Wang, L. (2019). A robust security architecture for SDN-based 5G networks. *Future Internet*, 11(4), 85.

5

Software-Defined Networking (SDN)

5.1. Introduction to Software-Defined Networking

Software-Defined Networking (SDN) is a software solution for automatically configuring a network. This solution introduces a separation of the control plane and the data plane. The control plane is centralized on a single machine called the controller. In short, SDN pushes for strong centralization of control and virtualization of SDN switches. Until now, flow tables were calculated in a distributed way in each router or switch device. In the new architecture, the calculations to determine the paths or routes are performed in the controller.

Traditional architectures are difficult to optimize because the distribution of calculations to obtain routing or switching tables leads to long latency times to allow control data to propagate throughout the network. In addition, networks are totally decoupled from services, and the configuration is complex to optimize all the requests. To simplify all this, SDN enables automatic configuration, very short start-up time because of virtualization and a configuration always adapted to the customer's demands, with the right quality of service and not a general quality of service.

The SDN architecture can be summarized in three basic principles. The first is the decoupling of the data and control planes: control packets are exchanged between the controller and the SDN nodes. The virtual devices are located in small, medium or large data centers. The second principle is to move the network from a physical form to a logical form. This new environment allows the network to be changed on the fly by adding, modifying or removing paths, routing tables or new networks at will. Finally, the third principle is the automation of operations conducted on the network, whether for management or control. This automation is achieved through centralization, since in the early 2020s we know how to use centralized artificial intelligence but not (yet) distributed artificial intelligence, even if progress allows us to hope for distributed control in a few years.

Storage, computing and networking use virtual machines associated with each of these domains that share physical resources. All three types of virtual machines or containers can be found on the same server. In order for the environment to run without failure, it is necessary to add virtual machines or containers for security and management and control. Today, these five groups of virtual machines need to be in place in a company to form an operational information system. The five domains are set up through virtual machines associated with each of the domains. A company's entire IT environment can therefore be concentrated in the cloud in the form of virtual machines, microservices or functions in data centers.

To the virtual machines of the digital infrastructure, we must add the application virtual machines, which can be of two types: business applications and applications for control or orchestration of the environment itself.

5.2. ONF architecture

Standardization is important for a control system to support the various devices that make up a company's information system. SDN standardization is the responsibility of the ONF (Open Networking Foundation), which was created under the aegis of major web and cloud companies, following the proposal of this new architecture by Stanford University and the company Nicira.

The architecture proposed by the ONF is described in Figure 5.1. It consists of three layers. The lowest layer is an abstraction layer that decouples the hardware from the software and is responsible for transporting the data. This layer describes the protocols and algorithms that allow IP packets or frames encapsulating IP packets to travel through the network to their destination. It is called the infrastructure layer. The second layer concerns control and gives rise to the control plane. This plane contains the controller that configures the data plane so that the routing of packets or frames is done in an optimal way. The vision of the ONF is to centralize the control to allow the recovery of numerous information coming from the clients and the logical equipment. This allows the centralized controller to realize Big Data from all the information collected in the controlled network. This Big Data is analyzed by artificial intelligence techniques. Of course, the reliability and security problems posed by centralized environments must be taken into account and therefore the machines involved in the control decisions must be duplicated or, more often, triplicated.

The controllers operate on a whole set of functions such as infrastructure provisioning or load balancing with regard to the various network equipment to optimize performance or reduce energy consumption. More specifically, the controller is in charge of configuring the equipment used in a network, such as routers, switches, firewalls, authentication servers and, more generally, all the servers required for the network to function properly. These different machines must be placed in the best possible locations to optimize the overall functioning of the network. This process is called the urbanization of virtual machines, microservices or functions on the company's or provider's data centers.

Finally, the top layer, or application plane, is responsible for the business applications needed by the clients and the storage, computation, network, security and management applications. This layer introduces the programmability of the applications and allows all the necessary information to be sent down to the controller to enable the opening of paths, routes or software networks that perfectly match the needs of the applications. This plane also includes the control, orchestration and management applications necessary for the life of a company's information system. The application plane must be capable of passing the information necessary to open the path, route or network corresponding to the application to the controller. Any new service can be introduced quickly and will give rise to a path, a new routing or a slice (a virtual network).

The three layers of the ONF architecture are described in Figure 5.1 and include application and their programmability, control with centralized intelligence, and abstraction at the infrastructure level. We will return to the interfaces between these layers which are of paramount importance for OEM products, enabling their compatibility. The ONF is continuing its standardization work, mainly on the intermediate layer and the interfaces. Some parts of the architecture are taken over by other standardization organizations to comply with legal standards.

The general architecture of the ONF is completely virtual. In order to return to a real architecture, an additional layer, the infrastructure layer, must be added to install the virtual environment. This additional layer is shown in Figure 5.2. We find the infrastructure layer which is here a digital infrastructure on which the virtual machines of the abstraction plane are positioned. The interface between the two planes is not standardized and leads to various implementations except in the Cisco architecture where the physical infrastructure must be specialized Cisco hardware.

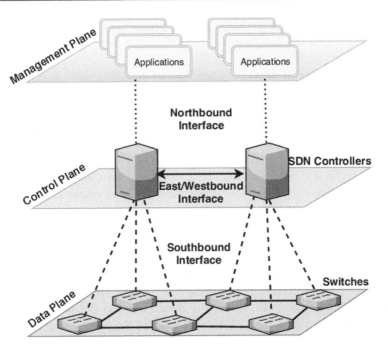

Figure 5.1. *The Open Network Foundation (ONF) architecture. For a color version of this figure, see www.iste.co.uk/haddadou/edge.zip*

This architecture requires data centers ranging in size from very small to very large depending on the size of the company and the distribution of resources in the Cloud Continuum. 5G telecom operators have not been wrong about the importance of investing in virtualization and setting up MEC (Multi-access Edge Computing) data centers to try to host virtual machines and more generally company applications. The major cloud manufacturers have also understood the importance of developing Edge data centers and are trying to find alliances with 5G telecom operators, especially to set up and manage their MEC data centers.

In the architecture described in Figure 5.3, we find the application, control and virtualization layers with the northbound and southbound application programming interfaces (APIs) between these layers, and the eastbound and westbound interfaces with the other controllers. The northbound interface allows communication between the application level and the controller. Its purpose is to describe the needs of the application and to pass commands to orchestrate the network. The northbound interface is a web type that we will examine later. The southbound interface describes the necessary signaling between the control plane and the virtualization layer. This requires the controller to be able to determine the elements that will form the path or routing to be implemented. In the other direction, network resource occupancy statistics must be fed back to the controller so that the controller has the most complete view of resource usage possible. The capacity required to retrieve the statistics may represent a few percent of the network's capacity, but this is essential for an optimization that will save much more than a few percent.

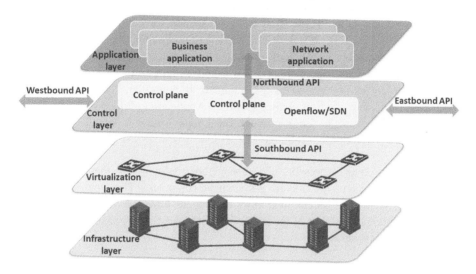

Figure 5.2. *SDN architecture. For a color version of this figure, see www.iste.co.uk/haddadou/edge.zip*

In addition to the two interfaces described above, there are the eastbound and westbound interfaces. The eastbound interface enables two controllers of the same type to communicate and make decisions together. The westbound interface must also allow communication between two controllers but belonging to different subnets. The two controllers can be homogeneous, but they can also be heterogeneous and in this case, a signaling gateway is necessary.

Figure 5.3 describes some important open-source software that has been developed to support a layer or interface. Starting at the bottom, in the virtualization layer, network virtual machines have been standardized by ETSI in a working group called

NFV (Network Functions Virtualization), which gave rise to the OPNFV platform that is now Anuket. NFV aims, as we have seen, to standardize all network functions to virtualize them and allow them to run as a virtual machine in a data center. To complete this standardization, an open-source code has also been developed to ensure compatibility between virtual machines supporting the same function.

The OpenDaylight controller, which we described in the previous chapter, was selected by Anuket/OPNFV as the controller to support the control of the platform. It has been developed as open source with the help of many companies. This controller contains a large number of modules often developed in the interest of the company that took over this work. OpenDaylight was mainly developed by Cisco and Red Hat but was not chosen by Cisco as its commercial controller. This controller is completely different and is called APIC (Application Policy Infrastructure Controller). There are of course many other open-source controllers such as OpenContrail, ONOS, FloodLight, etc.

The highest layer represents the management systems of the Cloud. It is somewhat equivalent to the operating system of computers. The OpenStack cloud manager has been the system that most developers have been working on and represents the largest number of the controllers, but many other products exist either in open source or proprietary.

The southbound interface is often known through its standard from the ONF, which is OpenFlow. OpenFlow is a signaling between the infrastructure and the controller. This protocol was designed by the company Nicira and has become a de facto standard through the ONF. OpenFlow carries the signaling that configures the nodes in the network and carries the statistics in the other direction.

Northbound and southbound interfaces have been standardized by the ONF to allow compatibility between cloud providers, control software and physical and virtual infrastructure. Most equipment vendors comply, but with varying degrees of difficulty, depending on the equipment they already have in operation.

Companies that make their choice on SDN have to adapt to the SDN environment and its infrastructure. This is possible because the software layer is normally independent of the physical layer. The company's machines must be compatible with the SDN vendor's hypervisor or container products. The company can implement an SDN environment on a part of its network and then increase it little by little. This solution involves a dual skill set of the old architecture and the new.

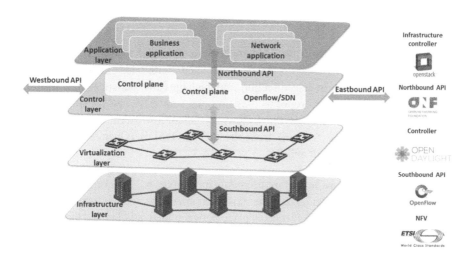

Figure 5.3. *The main open sources of SDN architecture. For a color version of this figure, see www.iste.co.uk/haddadou/edge.zip*

Now, we will go into the details of the different layers. Let us start with the lowest layer. The physical layer is designed around a digital architecture with data centers of different sizes, for example, the rise of open-source hardware (see Open Compute Project from the previous chapter). It is obvious that performance issues are crucial and the physical infrastructure must sometimes be up to the challenge. One of the priorities is to optimize the urbanization of virtual machines to avoid consuming too many energy resources. To do this, the best thing is to have an algorithm that enables the greatest possible number of servers to be put on standby outside of peak hours. Urbanization is becoming a keyword in these new generation networks. Unfortunately, urbanization algorithms are still very underdeveloped and they are not multi-criteria: only one criterion is considered, such as performance. The algorithm can be executed by taking load balancing as a criterion. But it can also take into account the energy expenditure by doing the opposite of load balancing, that is, by gathering the flows on common paths to switch off a maximum number of physical machines. The difficulty in the latter case is to turn resources back on as the load of the virtual machines increases. Some equipment such as virtual Wi-Fi access points are complex to wake up from standby when external terminals want to connect. First, electromagnetic sensors are needed to detect mobile terminals that want to connect and second, an Ethernet frame must be sent over a physical cable to the Wi-Fi access point with the Wake-on-LAN function.

5.3. Southbound interfaces and controllers

The southbound interface is located between the controller and the virtualization plane devices. The signaling protocol associated with this interface passes configuration commands in one direction and manages the statistical information feedback in the other. Many proposals exist in open source. A first list of protocols for this southbound interface is the following: OpenFlow from the ONF, I2RS (Interface to the Routing System) from the IETF, OvSDB (Open vSwitch Data Base), NetConf, SNMP, BGP and P4.

We will detail the most important protocols in Chapter 6. However, to introduce this southbound interface, let us take the most emblematic protocol: OpenFlow. This signaling is shown in Figure 5.4. It takes place between the controller and the network devices.

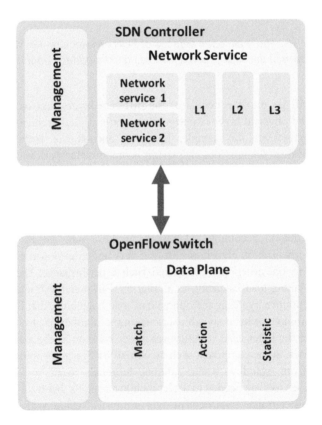

Figure 5.4. *The OpenFlow signaling protocol. For a color version of this figure, see www.iste.co.uk/haddadou/edge.zip*

The information transported follows the trilogy of correspondence, action and statistics, that is:

– flows can be uniquely determined by matching a number of fields corresponding to addresses, port numbers, identities, etc.;

– the actions transmitted from the controller to the network equipment can be specified, such as updating a routing or switching table or, more generally, a flow table;

– transfer statistics on port, transmission line and node utilization rates such as the number of bytes sent on each port can be specified.

This solution is standardized by the ONF, and we will describe it in more detail in Chapter 6.

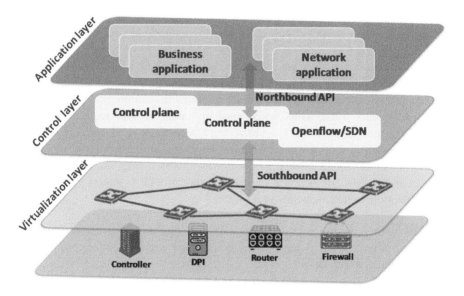

Figure 5.5. *The control level and its interfaces. For a color version of this figure, see www.iste.co.uk/haddadou/edge.zip*

The purpose of the controller is to configure the data plane and to receive from the application plane the elements necessary to determine the controls to be exercised. The position of the controller is shown in Figure 5.5. The controller receives requests from users via the application to be considered. From the application description, the controller must deduce the configuration to be established on the virtual network devices to obtain the appropriate quality of service and security. The actions are executed on the virtual machines: routers, switches, firewalls, load balancers, VPNs, etc. All that remains to be done is to position the virtual machines on the physical equipment necessary for the life of the network.

Many controllers have been developed in open source. OpenDaylight represents the best example since it has been chosen as the one to be used on the Anuket/OPNFV platform. OpenDaylight is free software from the Linux Foundation, as we discovered in the previous chapter. More than 50 companies have provided experienced developers to develop this controller. It has many configuration modules and various interfaces to the north, south, east and west.

The controller contains modules that take care of different functions necessary for the proper functioning of the network. One of the most important among these is a load balancer. In fact, this term indicates algorithms determining the best paths to follow in the data plane. This choice must optimize the user requests or, more precisely, the user applications. Load balancing is essentially valid during peak hours. During other hours, the load balancer must determine the best paths according to user desires. In fact, load balancing becomes load unbalancing: unlike during peak hours, as many flows as possible must be routed through common paths to shut down as many intermediate nodes as possible.

Based on the statistics received, OpenDaylight must decide which nodes in the network infrastructure the packets should pass through and pass on the configuration to optimize the applications.

5.4. The northbound interface and the application plan

The purpose of the northbound interface between the application plane and the controller is to transmit information from the applications so that the controller can open the best possible paths with the right quality of service, adequate security and the necessary management so that the operations can be carried out without problems. The basic protocol to carry out these transmissions is based on the REST (Representative State Transfer) API. This application interface allows sending the necessary information for configuration, operations and feedback. This protocol, based on RESTful, is integrated in many cloud management systems and in particular in the interfaces of most of these systems and in particular OpenStack in which the REST API is integrated. This interface can be found in Edge and Cloud Computing providers and finally in providers who offer a Virtual Private Cloud (VPC) API.

The representations allowed by the REST protocol enable information to be passed on the northbound interface with the following properties: each resource is uniquely identified, resources can be manipulated through representations, messages are self-describing: they explain their nature by themselves and each access to the following states of the application is described in the current message.

The application plane is essentially formed by virtual machines, microservices or functions in the Cloud Continuum. These machines can be of business type or network element management type such as handover management or determination of the best access for a multi-technology endpoint.

The standard is OpenStack, which we introduced in Chapter 3. It is a cloud management system, which has been chosen by many telecom operators and network providers, providing computing power, storage and network resources. OpenStack has already seen 26 releases in almost 13 years (from October 2010 to September 2022). The latest release of OpenStack, released in September 2022, is Zed. OpenStack is free software under the Apache license.

The architecture of OpenStack is described in Figure 5.6. It is modular and contains many modules such as for computation, Swift for storage, Glance for image service, Dashboard for setup and control console.

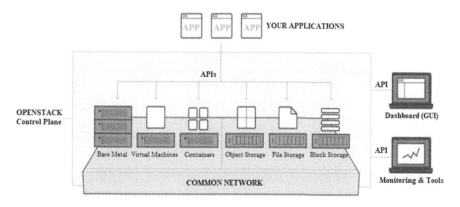

Figure 5.6. *The OpenStack management system (OpenStack). For a color version of this figure, see www.iste.co.uk/haddadou/edge.zip*

The part that interests us most in this book is Neutron, which deals with the networking module. In this framework, OpenStack provides flexible network models to support applications. In particular, OpenStack Neutron manages IP addresses and allows static addresses or uses DHCP. Users can create their own network in which SDN technology is supported. OpenStack has many extensions such as IDS (Intrusion Detection Systems) intrusion detection, load balancing, the ability to deploy firewalls and VPNs.

5.5. Conclusion

To conclude this section and summarize SDN architectures, we have seen that there are different components that have been put in place to achieve a global operation. The upper and lower parts concern the Cloud and the physical and logical networks. Between these two levels, the management and control of the network and applications must be put in place. On the business application side, there are the sets of software modules, mostly open source, that enable the deployment of Cloud computing infrastructures and more precisely of IaaS (infrastructure as a Service). On the other side, we find the applications for setting up virtualized network structures with the necessary commands to support business applications. Finally, Figure 5.7 shows the interconnection of controllers to achieve end-to-end configuration. The SDNi interface has been standard since 2021.

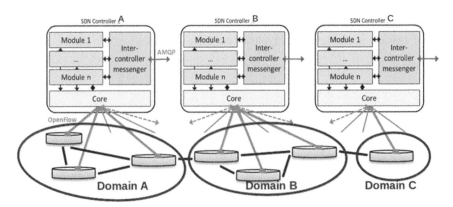

Figure 5.7. *The interconnection architecture of different SDN domains. For a color version of this figure, see www.iste.co.uk/haddadou/edge.zip*

5.6. References

Abigail, J., Segun, I., Popoola, A., Aderemi, A. (2018). Software-defined networking: Current trends, challenges and future directions. In *International Conference on Industrial Engineering and Operation Management*, Washington, DC.

Azodolmolky, S. (2013). *Software Defined Networking with OpenFlow*. Packt Publishing.

Blanco, B., Fajardo, J.O., Giannoulakis, I., Kafetzakis, E., Peng, S., Perez-Romerof, J., Trajkovskag, I., Khodashenash, P.S., Gorattii, L., Paolinoj, M., Sfakianakis, E. (2017). Technology pillars in the architecture of future 5G mobile networks: NFV, MEC and SDN. *Comput. Standards Interface*, 54, 216–228.

Chayapathi, R. and Hassan S.F. (2016). *Network Functions Virtualization (NFV) with a Touch of SDN*. Addison-Wesley.

Chen, W., Wang, D., Li, K. (2019). Multi-user multi-task computation offloading in green mobile Edge Cloud Computing. *IEEE Trans. Serv. Computer*, 12, 726–738.

Goransson, P. and Black, C. (2014). *Software Defined Networks. A Comprehensive Approach*. Morgan Kaufmann.

Hu, F. (2014). *Network Innovation through OpenFlow and SDN: Principles and Design*. CRC Press.

Iqbal, M., Iqbal, F., Mohsin, F., Rizwan, M., Ahmad, F. (2019). Security issues in software defined networking (SDN): Risks, challenges, potential solutions. *International Journal of Advanced Computer Science and Applications*, 10(10), 298–303.

Jarraya, Y., Madi, T., Debbabi, M. (2014). A survey and a layered taxonomy of software-defined networking. *IEEE Communications Surveys & Tutorials*, 16, 1955–1980.

Morreale, P.A. and Anderson, J.M. (2014). *Software Defined Networking: Design and Deployment*. CRC Press.

Nadeau, T.D. and Gray, K. (2013). *SDN: Software Defined Networks*. O'Reilly.

Priyadarsini, M. and Bera, P. (2019). A secure virtual controller for traffic management in SDN. *IEEE Lett. Comput. Soc.*, 2(3), 24–27.

Rana, D.S., Dhondiyal, S.A., Chamoli, S.K. (2019). Software defined networking (SDN) challenges, issues and solutions. *International Journal of Computer Science and Engineering*, 7(1), 884–889.

Rawat, D.B. and Reddy, S.R. (2017). Software defined networking architecture, security and energy efficiency: A survey. *IEEE Commun. Surv.*, 19, 325–346.

Shukla, V. (2013). *Introduction to Software Defined Networking: OpenFlow & VxLAN*. Createspace Independent Pub, Kindle Store.

Stallings, W. (2015). *Foundations of Modern Networking: SDN, NFV, QoE, IoT, and Cloud.* Addison-Wesley.

Xia, W., Wen, Y., Foh, C.H., Niyato, D., Xie, H. (2015). A survey on software-defined networking. *IEEE Commun. Surv. Tutor.*, 17, 27–51.

6

Edge and Cloud Networking Commercial Products

6.1. Introduction to SDN products

In the previous chapters, we introduced SDN, its centralized architecture and its use as virtual machines. In this chapter, we will examine the main categories of SDN-related products. They can be classified into six main classes. The first class is the fabrics that form the access networks to the servers in data centers. SD-WANs (Software-Defined Wide Area Networks) represent the second major class and certainly the most important economically. They concern a solution enabling the optimization of the use of a company's WAN networks. The third class is the virtualization of the local area network under the name of vCPE (Virtual Customer Premises Equipment). The fourth category is antenna box virtualization, especially Wi-Fi (vWi-Fi). The fifth category contains the vRAN (virtual Radio Access Network) which proposes the virtualization of the access network between the antennas on which the users are connected and the core network. Finally, the last class concerns the virtualization of the core networks of 4G and 5G: vEPC (virtual Evolved Packet Core) and 5GC (5G Core). We will describe these different classes in the following sections.

6.2. Fabric control

A fabric is the network associated with a data center that allows us to go from one server to another, internally. The fabric is also used to access any server from the outside. The fabric has horizontal and vertical rows offering multiple paths to get from one server to another. The objective of fabric control is to achieve load balancing and optimal traffic distribution on these paths. The SDN solution proposes to centralize in a controller the commands to distribute the flows on all the paths of the fabric. Two fabrics are described in Figure 6.1: the old generation and the new one used in the SDN framework.

The figure on the left represents the still frequently used solution consisting of one tier of switches and one tier of routers. The new generation has two tiers of switches to achieve even greater capacity through the speed of the switches. The switches closest to the servers are called leaf switches, while the switches closest to the output are called spine switches. These switches allow any server to be connected to any other server through multiple paths within the data center. These paths are provisioned and set up by the SDN controller in a fully automatic way to optimize the total throughput. It is possible to replace switches with routers by performing Layer 2 routing. This is, for example, the case with the Cisco fabric. The advantage in this case is that multiple routes can be used simultaneously. IS-IS routing is generally used at level 2 with specific level 2 addresses.

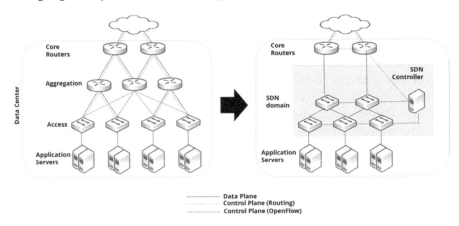

Figure 6.1. *Architecture of an old and new generation fabric. For a color version of this figure, see www.iste.co.uk/haddadou/edge.zip*

The SDN controller can also take care of controlling several fabrics simultaneously by managing intra and inter data center flows. Figure 6.2 shows such a case with two fabrics interconnected by a WAN (Wide Area Network). The protocols used in intra- and inter-fabric communications will be presented in a specific chapter devoted to Cloud Networking protocols.

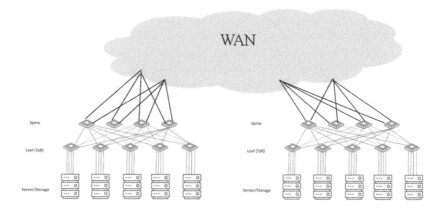

Figure 6.2. *Interconnection of fabrics. For a color version of this figure, see www.iste.co.uk/haddadou/edge.zip*

Many architectures have been commercialized since 2015 to support fabric control. We will describe some of them.

6.2.1. *NSX from VMware*

VMware was the first major company to embrace virtualization. It is a software company, not a hardware company, which is why its NSX architecture is entirely software and relies on hardware from network equipment vendors and cloud providers. The NSX architecture is shown in Figure 6.3. This architecture is entirely software and highly open by taking over open-source software machines. VMware's platform is essentially based on the virtualization of the NSX architecture. This platform contains the main networking and control elements. On the network side, the machines can be routers or Layer 2 or Layer 3 switches, depending on the user's needs. On the control side, the platform contains the basic elements with the ability to position virtual firewalls, manage flows to transit through paths determined by the load balancing machine and by opening virtualized VLANs (Virtual LANs). The word virtual is used twice in virtualized VLAN: the VLAN is a software-based local network that uses physical machines that are not necessarily located in the same place, and this local network is itself located in a virtualized universe.

The top layer is formed by the cloud management systems which can be either VMware or from open-source cloud management systems such as OpenStack, which is the standard used in the OPNFV-based architecture. The *northbound* interface is REST based. The software network part relies either on data centers with internal virtual equipment or on physical machines from different equipment manufacturers with which VMware has developed partnerships. The protocol for the *southbound* interface is OpenFlow, developed by Nicira, a company acquired by VMware for more than a billion dollars. The *southbound* interface can however adopt other techniques such as the Open vSwitch interface: OvSDB (Open vSwitch Data Base).

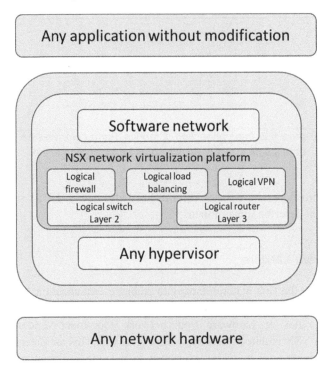

Figure 6.3. *NSX architecture. For a color version of this figure, see www.iste.co.uk/haddadou/edge.zip*

The hypervisors or containers to support the virtual machines can be of any type as well as the physical hardware. The basic virtual appliance is Open vSwitch from the Linux Foundation, which is a very advanced switch with almost all the possible properties of a switch. Figure 6.4 shows the NSX architecture with its various components in more detail.

From a security point of view, the switch has VLAN isolation and traffic filtering. Monitoring is done through different protocols such as NetFlow or specific versions from VMware related vendors. Quality of service is supported by traffic shaping and traffic filtering techniques. The *southbound* interface to the controller uses several protocols including OpenFlow, which is the preferred protocol, but also OvSDB and various network machine configuration protocols.

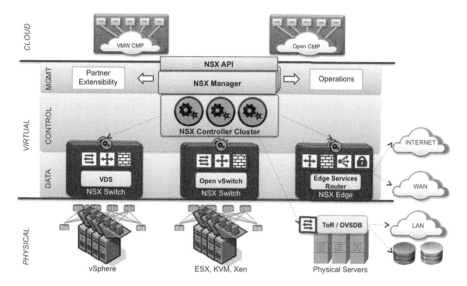

Figure 6.4. *The detailed architecture of NSX (VMware). For a color version of this figure, see www.iste.co.uk/haddadou/edge.zip*

The NSX Controller's logical firewall has opted for a distributed, virtualization-oriented firewall with activity identification and monitoring. It uses micro-segmentation, which is a security architecture that breaks the data center into distinct logical security sections down to the individual workload level. We can even talk about femto segmentation when we go down to the workloads linked to the functions that will replace the microservices little by little. Security controls are exercised on single and increasingly smaller segments. Micro-segmentation enables IT to deploy flexible security policies deep in the data center by leveraging network virtualization technology instead of installing multiple physical firewalls. In addition, micro-segmentation can be used to protect each virtual machine or container in a company network with security controls based on predefined application-level policies. Because security rules are applied to individual workloads, micro-segmentation software can greatly improve a company's resistance to cyberattacks by only allowing attacks on small areas.

The way NSX works allows for complete dissociation of the physical infrastructure with the paths formed in the fabric. Virtualization allows them to be overlaid on any physical hardware with input/output cards adapted to the transmissions. In addition, NSX works with any type of hypervisor to make servers easily virtualizable. The NSX system allows the replication of the physical network model in software form whether it is Layers 2, 3, 4 or 7. In addition, NSX provides automation through a RESTful (REpresentational State Transfer) API that allows cloud management platforms to automate the provisioning of network services. Finally, NSX provides a platform to insert services from other vendors. These software and hardware products from VMware's partner physical machine integration include network gateway, application provisioning and security services.

6.2.2. *Cisco Application Centric Infrastructure*

Cisco was quite slow to determine its architecture vis-à-vis SDN. It was achieved in 2014. This architecture sees applications placed at the center of the game. The goal is to automate the Layer 2 routing tables to optimize the transport of frames on the fabric using only physical Cisco machines. This architecture is shown in Figure 6.5. It is, as its name indicates, application centric since the routes depend on the characteristics of the service to be performed. However, it is also a hardware-based architecture since it requires Cisco hardware to be able to implement it, even if physical machines from other manufacturers can complete the hardware part.

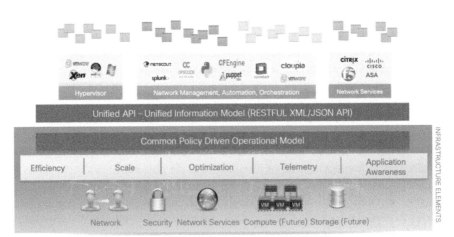

Figure 6.5. *Cisco's ACI architecture (Cisco). For a color version of this figure, see www.iste.co.uk/haddadou/edge.zip*

Figure 6.5 shows Application Centric Infrastructure (ACI), which starts with the applications and goes to the controller, which has numerous functions for configuring the routers and, by extension, all the equipment constituting the network (switch, router, firewall, DPI [Deep Packet Inspection], etc.). The northbound interface in particular is developed at Cisco. It uses a REST representation like most SDN products to transport high-level business rules from the application layer to the controller which are transformed into network rules. The cloud aggregates the applications and addresses through an orchestrator, for example, the OpenStack orchestrator, to the controller through a Cisco API. The controller has many intelligent modules to interpret the business rules from the northbound interface and transform them into network rules, using for example DiffServ, which are then sent through the southbound interface to configure the network nodes.

The controller of this architecture is called APIC (Application Policy Infrastructure Controller). It is at the heart of Cisco's architecture. It uses artificial intelligence techniques to determine the relationships between application needs and the networks that must support them. Since 2020, Cisco has used a new solution to optimize the configuration. This new generation comes from Intend-based technology, that is, based on the users' intention. Each user must tell the controller what their flow intentions are. From the intentions of all users, the controller deduces an optimal configuration that must be able to handle all flows and give them the right quality of service. In fact, compared to the classic version, this technology avoids having to reconfigure at too high a rate, which ends up not being optimal due to a large overhead.

The complete architecture of ACI is described in Figure 6.6. It can be seen that it is based on a physical infrastructure from Cisco through Nexus 9000 machines. The southbound interface protocol is OpFlex, specific to Cisco, which is not at all compatible with OpenFlow since it carries network policy rules. OpFlex signaling has however been submitted to the IETF and has become an RFC but almost no manufacturer has adopted it outside of Cisco. It is an open-source protocol that we will study in more detail in Chapter 7.

The protocol used in the Cisco fabric is a TRILL type routing that we will detail in Chapter 9.

The components of ACI architecture are:

– Cisco's APIC, which is the main component of the architecture;

– application chaining profiles: the application chaining profile corresponds to the logical representation of all the components necessary to realize a service and their interdependence on the hardware (fabric);

– the physical equipment (fabrics) required for the Cisco ACI architecture ; these belong to the Cisco Nexus range. Cisco Nexus 9000 machines operate either in Cisco NX-OS mode for compatibility and consistency with older Cisco Nexus switches, or in Cisco ACI mode to take full advantage of policy-based services and infrastructure automation capabilities.

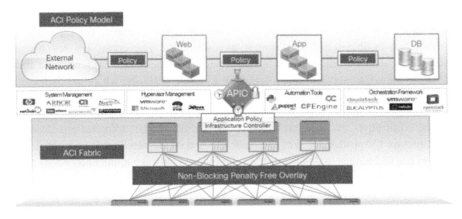

Figure 6.6. *The detailed architecture of ACI (Cisco). For a color version of this figure, see www.iste.co.uk/haddadou/edge.zip*

6.2.3. *OpenContrail and Juniper*

Juniper also released its own SDN architecture in 2014 based primarily on open-source software but with Juniper-specific additions. The center of this architecture is the controller which for a long time was OpenContrail. In the late 2010s, two branches split Tungsten Fabric, which continued to develop in open source, and Contrail, which became Juniper's specific controller. Juniper's Contrail is an open, simple and agile SDN solution that automates and orchestrates the creation of virtual networks on demand.

Virtual networks are perfectly suited to applications that require them to be open. We can highlight the simplicity of this architecture for the creation of virtual networks that integrate seamlessly with existing physical networks and that are simple to manage and orchestrate. The environment is also open. There is no dependence on any particular vendor and it eliminates the costs associated with OEM (Original Equipment Manufacturer) through an open architecture. Juniper's platform works with a wide range of open hypervisors, orchestrators and physical networks. Finally, a huge benefit is the solution's agility, which reduces time-to-market for new services by automating the creation of virtual networks to interconnect private, public and hybrid clouds.

Service providers can use Contrail to deliver a range of innovative new services, including cloud-based offerings and virtualized services. Juniper's platform is depicted in Figure 6.7. Note in this figure that the platform primarily uses OpenStack, but this is not a requirement; other cloud management software is also acceptable.

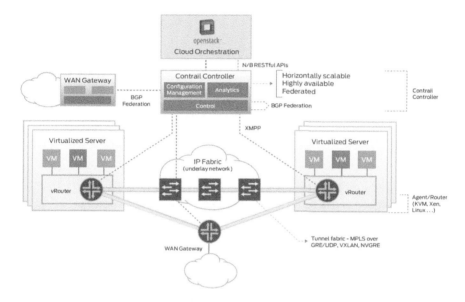

Figure 6.7. *Juniper's OpenContrail Platform (Juniper). For a color version of this figure, see www.iste.co.uk/haddadou/edge.zip*

6.2.4. Nokia SDN Architecture

Nokia's solution comes largely from its subsidiary Nuage Networks. The architecture is described in Figure 6.8. It follows the main lines already described above by other equipment manufacturers. The part where we can find a strong specificity from Nokia concerns optical and mobile networks. In particular, Nokia offers a vEPC (virtual Enhanced Packet Core), that is, a core network for fourth generation mobile networks. In its products, Nokia favors the virtualization of functions and is involved in having a strong impact on the standardization of NFV.

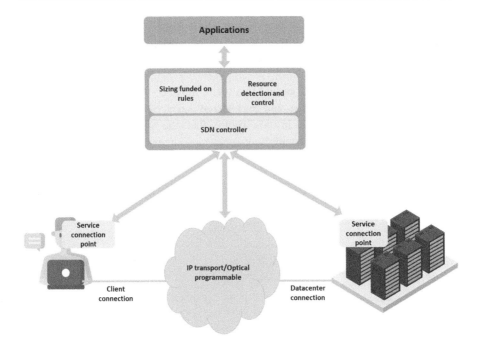

Figure 6.8. *The Nokia SDN platform (Nokia). For a color version of this figure, see www.iste.co.uk/haddadou/edge.zip*

6.3. Software-Defined Wide Area Network

SD-WAN is a solution to control and manage the multiple WANs of a company. There are various types of WANs: Internet, MPLS, 4G LTE, 5G NR, DSL, fiber, cable network, circuit link, etc. SD-WAN uses SDN technology to control the entire environment. As with SDN, there is a separation of the data plane from the control plane. A centralized controller must be added to manage flows, routing or switching policies, packet priority, network policies, security, etc. SD-WAN technology is based on an overlay, that is, nodes that represent underlying networks.

6.3.1. *The basics of SD-WAN*

Figure 6.9 shows an SD-WAN. It shows a set of WANs, represented by three networks: an MPLS network, generally set up by a telecom operator or integrator; the Internet network, which is an extremely inexpensive or even free network, but without knowing the quality of service at all times; and finally a telecom operator's network, represented by its core network, which is accessed via a 3G, 4G or 5G connection. The choice between these three networks is not simple, because it depends on the cost, the quality of service required by the application, the time of day, the security level, etc. The objective of SD-WAN is to direct packet flows to the right networks to optimize performance criteria that generally include cost, quality of service and security. The main idea is to use the cheapest network, that is, the Internet, as soon as possible to lower the cost of company communications.

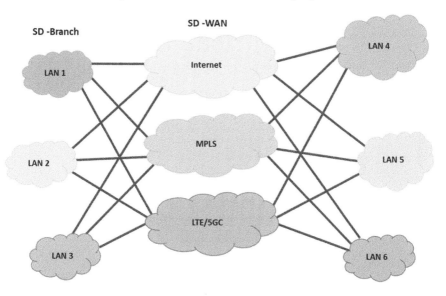

Figure 6.9. *An SD-WAN network. For a color version of this figure, see www.iste.co.uk/haddadou/edge.zip*

We will take a look through the following figures to see how the SD-WAN works. An example of a company network is shown in Figure 6.10. The company has several networks, and in particular several WANs, to carry out communications between its various campuses and branches, the head office and the data centers. We can add the access networks to the WANs. We can even go further by integrating the local networks to take into account the end-to-end. In the latter case, the solution is called SD-Branch to indicate that the company's branches are considered.

From the network topology, the overlay network must be built, which has been done in Figure 6.11. To do this, we must add network equipment based on probes capable of measuring performance with neighboring probes. The main performances correspond to the response time between two probes located respectively at the entry and exit of a network, the flow, the security, etc. In other words, the probes must measure the performance of the networks that make up the company network.

98 Cloud and Edge Networking

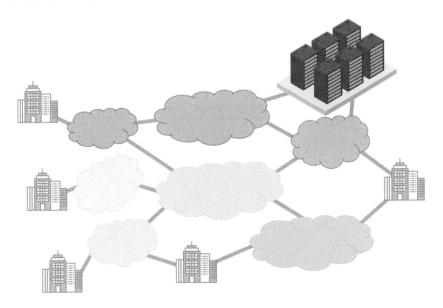

Figure 6.10. *Company networks. For a color version of this figure, see www.iste.co.uk/haddadou/edge.zip*

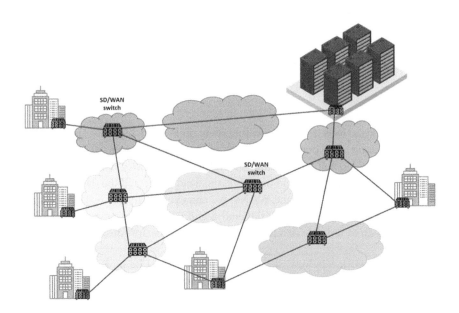

Figure 6.11. *The SD-WAN overlay network. For a color version of this figure, see www.iste.co.uk/haddadou/edge.zip*

Once the performance of the different networks is obtained, these networks can be removed to leave only the overlay network, which represents the overall network and the associated measurements to get from one probe to another. Figure 6.12 shows the overlay network where only the intermediate equipment that separates the company networks remains.

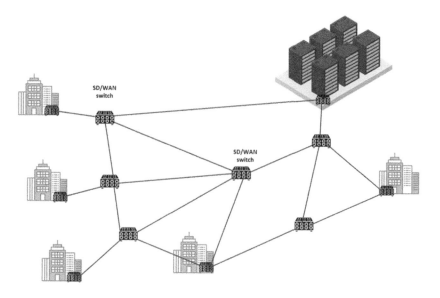

Figure 6.12. *The overlay network. For a color version of this figure, see www.iste.co.uk/haddadou/edge.zip*

The next step is to send all the measurements to a controller, which is responsible for automating the decisions regarding which networks to use to obtain the quality of service required by the application. Figure 6.13 describes this architecture with the controller. The controller can be located in different places, such as the company's headquarters, the company's data center, a node in the center of the network, a server in the MPLS network or a data center owned by an integrator or service provider.

The key elements of SD-WAN are described in the following points.

– Optimization of the choice of the WAN through which the packets must transit to optimize the cost of the network traversal.

– The use of probes integrating DPI to determine the response times and throughputs of each type of flow, in each network.

– The ability to switch flows by application, by quality of service and by security features.

– Optimization of access to the company data centers for all flows from all SD-WAN access points.

– Communication security by opening secure channels, end-to-end or on specific networks.

– Automation of the configurations thanks to the controller.

– Improving availability by allowing multiple connections between two points on the network.

– Consideration of the security of each WAN.

– Implementation of security on all protocol layers.

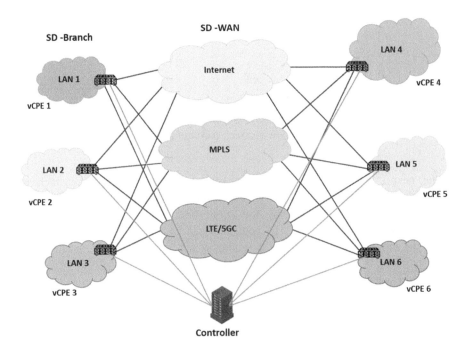

Figure 6.13. *The SD-WAN with its controller. For a color version of this figure, see www.iste.co.uk/haddadou/edge.zip*

Depending on the controller's position, three approaches to SD-WAN can be defined. The first approach, CPE centric, involves a controller positioned in one of the company's local networks, accompanied by numerous virtual machines also located in the user's local network. The second approach, WAN centric, involves the use of virtual machines located in one or more data centers located in the WAN. The third possibility is a hybrid approach in which some virtual machines are in the CPE and others in the WAN. For example, in the CPE, there may be a WAN optimizer, cache servers, NAC (Network Access Control), etc., and in the WAN, an IPS (Intrusion Prevention System), a malware detection system, etc.

SD-WAN players include several industry categories including:

– Publishers who resell their software to large accounts and especially to integrators.

– Telecom equipment manufacturers who offer customers all the software and hardware required to implement an SD-WAN.

– Integrators who offer a complete solution based on software from publishers and hardware from telecom equipment manufacturers. Telecommunications operators are often part of the integrators.

– SD-WAN operators who have all the elements to implement a complete solution.

6.3.2. *SD-WAN 2.0*

SD-WAN 2.0 represents the second generation. Its definition is not precise, but essentially concerns extensions. The first one we will study in the next section is the SD-Branch, that is to say, the addition of company branches that we can translate into the local networks of the company's different sites. A branch's network can be very simple but sometimes complex. This network can be virtualized in vCPE or not. It is a question of completing the values of parameters complementary to those of the WAN. The values of these parameters can significantly modify the averages and variances of the response times by considering the end-to-end, that is, from the sender of a packet to the terminal receiver of this same packet. The parameter values can come from a local controller or from measurements from sensors at the ends of the network.

The second feature introduced in SDN 2.0 is intelligent routing of SD-WAN switches. The terms seem contradictory since a switch switches and does not route. More accurately, SD-WAN devices are switches. In fact, the controller's objective is to set up switching paths and the SD-WAN switches only switch frames on the paths defined by the controller. The objective called intelligent routing is to change the path according to the network parameters: as soon as there is an acceptable and cheaper path, a rerouting is performed. This rerouting is called a routing to indicate that it can potentially be done for each frame received by the SD-WAN switch. In the latter case we have a level-two routing, and the controller must be able to introduce a new table for each frame. Since this is impossible, it is indeed a rerouting and not a routing. This term is used to indicate rapid changes in the switching tables. Intelligent routing generally uses artificial intelligence tools such as big data and machine learning and takes more and more parameters into account, especially security.

The third major feature concerns security, which has been added by companies specializing in this area, such as Fortinet and Palo Alto, which are now among the top five SD-WAN vendors. The main security solution for the company with an SD-WAN is called SASE (Secure Access Service Edge). It contains many security features that are heavily based on not trusting anyone, even the company president. We will describe these various extensions in the following.

6.3.3. *SD-Branch*

The SD-Branch includes all networks within the company (LAN, Wi-Fi, etc.), that is, both WAN and LAN networks. Figure 6.14 gives a first view of an SD-Branch network. In the solution proposed in this figure, the local networks have a vCPE infrastructure, and more precisely a uCPE (universal Customer Premises Equipment) where the u indicates user. A fog data center per site of the company groups the virtual machines useful to each site. It also retrieves all the values of the parameters necessary for the implementation of a Software-Defined Networking technology. A central controller retrieves all the values of all the LAN and WAN parameters and decides on the best paths to implement.

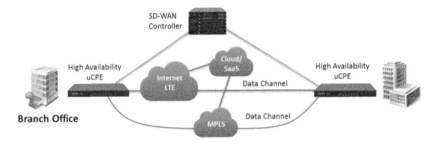

Figure 6.14. *An SD-Branch network. For a color version of this figure, see www.iste.co.uk/haddadou/edge.zip*

6.4. Secure Access Service Edge

SASE (Secure Access Service Edge) is a set of security features for the WAN that provides solutions to identify malware or malicious activity. SASE offers a set of independent virtualized functions that can be added rather than a general package.

The term SASE refers to an environment that sits on the Edge and allows secure access in SD-WANs. The end result is a flexible network architecture that optimizes both performance and security. In particular, this solution provides adequate performance for applications hosted on public, private and hybrid clouds as well as high-level security through a number of specialized features.

All SD-WAN vendors have introduced SASE solutions into their product suites, with a focus on simplifying Edge modernization and transforming the network to meet the demands of cloud computing.

A SASE solution can have many components that vary quite a bit from vendor to vendor and also depend on the history of the solution provider. But as a general rule, every SASE architecture provides the following five components:

– SD-WAN: SASE is integrated into SD-WAN technologies, that is, the integration of a global controller that recovers all the performances of the different WANs to allow a calculation of the best path taking into account the path cost.

– CASB (Cloud Access Security Broker): CASBs are on-premises or cloud-based security policy enforcement points. These enforcement points are placed between cloud service consumers and cloud service providers. They act on company security policies when accessing cloud resources.

– ZTNA (Zero Trust Network Access): ZTNA is a product or service that creates a logical access boundary based on identity and context around an application or set of applications. Applications provide a topology to avoid discovery, and access is limited, via a trusted third party, to a set of well-defined entities.

– FWaaS (Firewall as a Service): SASI firewalls protect the architecture against cyberattacks through tiered filtering and threat detection techniques.

– SWG (Secure Web Gateway): These gateways filter and prevent data breaches on the network by applying additional web filtering and access control measures.

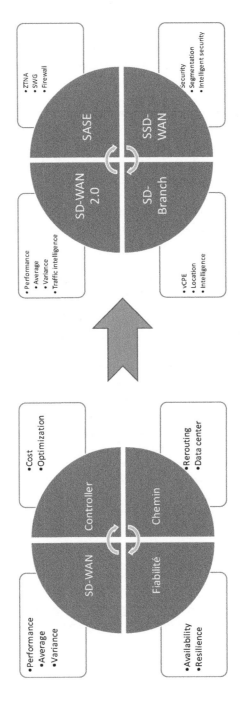

Figure 6.15. *An SD-WAN with SASE. For a color version of this figure, see www.iste.co.uk/haddadou/edge.zip*

SASE configurations can be classified into two main categories:

– Self-hosted/self-managed: this SASE configuration is designed and managed by the end user. The configuration includes a vendor-supplied SD-WAN environment and built-in security features.

– Vendor managed: this SASI configuration contains a vendor's platform (e.g. zScaler, Palo Alto Prisma, or Netskope) with the SASE architecture. The customer is then simply responsible for maintaining the connection to the platform.

The biggest advantage of a SASE architecture comes from a strong improvement of security for companies. SASE's Zero Trust security model requires identity verification for any user and protects data both during migration and on their endpoint. In addition, FWaaS and SWG features help protect users from different types of cyberattacks.

In addition to these intrinsic features, SASE gives users the ability to customize their architecture with other security properties such as web application and API protection, and inline encryption/decryption, to introduce just two. Consistent application of security policies makes network management easier than ever, without sacrificing the depth of functionality provided by SASI.

By replacing multiple on-premises appliances and infrastructure with one software stack, investment, operating, transport and asset costs can be significantly reduced. In addition, network changes and upgrades are much simpler with an integrated SASE architecture than in a WAN with independent, separate components.

SASE's centralized network management by the controller and consistent policy enforcement make it easy to adapt to traffic fluctuations for a reliable user experience. SASE also allows users to easily scale their network as they grow and technology upgrades, while protecting their configuration and maintaining optimized performance over time.

6.5. Virtual Customer Premises Equipment

vCPE concerns the virtualization of the local network. This product could have been called a vLAN since it is about virtualizing the local network. But since this acronym is already in use, we had to find another one. The CPE itself includes the equipment that is located on the company's premises, such as routers, switches, firewalls, intermediate equipment such as middleboxes. These devices are virtualized in the vCPE and the main issue is to determine where to place these virtual machines. The choices are similar to those made for the controller in the SD-WAN. The LAN virtual machines should be positioned in a data center that is either in the LAN, the operator's PoP (Point of Presence) that manages the LAN access interface, an operator data center, or a company data center that uses the operator's network for access.

A particular case called uCPE corresponds to the implementation of virtual machines in a white box which is a data center using open-source hardware. This free hardware, whose specifications are open, can for example come from the OCP (Open Compute Project). The white box contains virtual machines of type VNF (Virtual Network Function) which realize the functions of the CPE. An illustration of the vCPE is available in Figure 6.16 in which the main virtual machines are positioned in a hybrid way in the local network and in the core network.

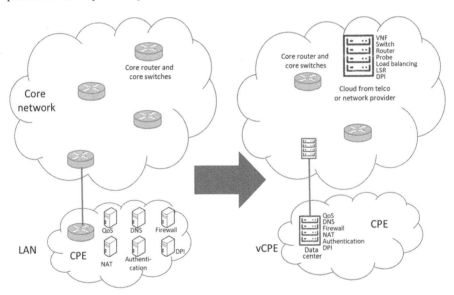

Figure 6.16. *An example of vCPE with a hybrid architecture. For a color version of this figure, see www.iste.co.uk/haddadou/edge.zip*

Increasingly, the two products SD-WAN and vCPE are being integrated into a single product called the SD-Branch. The virtual machines that realize the CPE are supported by the SD-Branch controller. As a result, a service can be opened taking into account both local and WAN networks. This enables managing the end-to-end quality of service as well as the security provided by SASE.

6.6. vWi-Fi

The virtualization of Wi-Fi can be seen as a special case of the virtualization of all the boxes associated with antennas. We will see in the next section the case of 5G antennas and the virtualization of the electronic cabinets that are located on the antenna mast or at its foot. The vWi-Fi can take several aspects: a standard virtualization in which the functions that are in the box are moved to a data center that is in the CPE or further away. If the virtual machines are far away, one must beware of the latency time which can grow excessively. A second possibility for virtualization is to put several access points in the same box. These different access points share the physical antenna. The advantage of having several SSIDs (Service Set Identifier) in one access point is that we can have specific properties associated with each access point.

We will start with the standard case of virtualization. In the case of complete virtualization, the box disappears completely including the signal processing. In this case, it is necessary to be able to connect the antennas directly to the data center which performs the signal processing to retrieve the binaries. Thus, the antennas have to be connected with a RoF (Radio over Fiber) technology, which transports the signal coming from the users and arriving at the antenna, to the data center. The cabling is therefore quite different from what exists today in companies that are essentially wired to the Ethernet. We will find this solution to connect the 5G antennas to the data center. This configuration is shown in Figure 6.17.

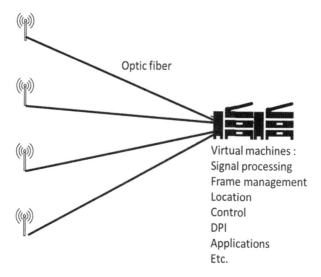

Figure 6.17. *The basic configuration for vWi-Fi*

In this configuration, all functions are located in the data center. We have signal processing, packet and frame management, geolocation and control as well as applications that are not found in the access point such as DPI and, directly, applications that can of course be multiplied.

However, this first solution is generally not compatible with the Ethernet network that may be available on the premises. This is the reason for the more usual case, shown in Figure 6.18, where the antennas are accompanied by a box with the basic functions to connect directly to the Ethernet network. In this case, the box has to process the signal and convert it into Ethernet frames, so that it can connect directly to the infrastructure network, that is, the available Ethernet network. The data center can receive all the additional functions such as a DPI, a firewall or an authentication server.

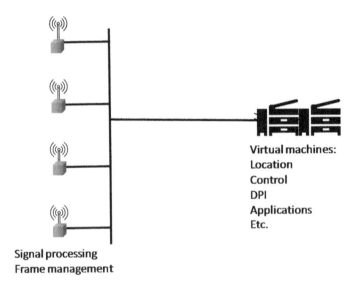

Figure 6.18. *vWi-Fi in Ethernet compatible mode. For a color version of this figure, see www.iste.co.uk/haddadou/edge.zip*

The second virtualization solution where the box holds the antenna is instead adapted to a set of operators sharing a Wi-Fi antenna. Each user has their own configuration and is independent of the other operators, which would be impossible in the case of different SSIDs. This virtualization can also be adapted to an HNB (Home Node B) box, that is, the equivalent of a 4G or 5G antenna but on a household level. The range of HNBs is generally quite similar to that of Wi-Fi equipment, although the maximum power can be higher so that the range can extend beyond the home and cover part of the street. This allows outside clients to connect to another user's HNB if it has not been restricted to the home where it is located.

6.7. Virtual Radio Access Network

The vRAN corresponds to the virtualization of the RAN equipment, that is, the network that connects the 4G or 5G antennas to the core network. This equipment essentially contains the RRH (Remote Radio Head) which takes care of the conversion of the radio signal into a digital signal and then the transformation of this digital signal into an optical signal to send the digital signals to the BBU (BaseBand Unit). The RoF technology is used to transmit the digital signal. Then, the BBU can connect to the core network to transmit the packets to the recipient. The BBU performs the processing of the digital signal from the RRH. Many solutions for bundling functions can be implemented. Figure 6.19 shows the solution proposed in the C-RAN (Cloud RAN). In this case, the BBU equipment is virtualized and moved to the operator's data center, which in the case of 5G is less than 10 km away, although some equipment manufacturers are pushing the distance between the antenna and the data center to 50 km.

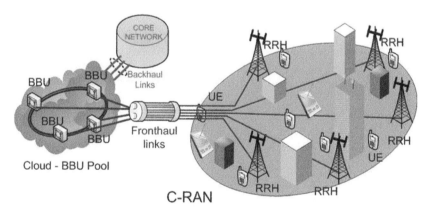

RRH : Remote Radio Head
BBU : Baseband Unit
UE : User Equipment

Figure 6.19. *Cloud-RAN (C-RAN): a 5G vRAN. For a color version of this figure, see www.iste.co.uk/haddadou/edge.zip*

Several antennas have their BBUs multiplexed in the same data center to perform load balancing and optimize resource utilization. The data centers of 5G operators called MEC (Multi-access Edge Computing) are interconnected by a loop to also perform load balancing between data centers.

vRAN is an application of NFV at the radio access network (RAN) level where all the access network equipment is virtualized. It is a centralization and virtualization of the hardware part of the base stations in the cloud. The MEC data centers are close enough to the antenna to allow a latency time of about 1 ms corresponding to the round-trip time of the user to the data center.

6.8. Virtual Evolved Packet Core and virtual 5GCore

vEPC and v5GC refer to the virtualization of the core network of 4G and 5G mobile networks. This virtualization concerns all the equipment required to build the core network. They are virtualized in one or more data centers depending on the size of the network, the number of data centers and the expected multiplexing.

In the vEPC, there are devices such as SGs (Serving Gateways) that route packets from access networks, PDNGs (Packet Data Node Gateways) that interface between the core network and other data networks, PCRFs (Policy and Charging Rules Functions) that handle the client interface.

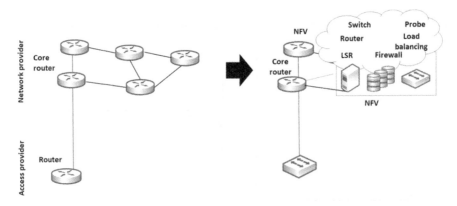

Figure 6.20. *Virtualization of a core network. For a color version of this figure, see www.iste.co.uk/haddadou/edge.zip*

In the v5GC, there are quite identical virtualized 5G control functions and 5G applications that are defined in release 17 from June 2022. The virtualization is achieved using the MEC data centers that are up to 10 km away from the 5G antenna.

More generally, the core network is virtualized by positioning virtual machines in one or more data centers either in the RAN or in the core network. The virtualized equipment includes, as shown in Figure 6.20, routers, switches, Label Switch Routers (LSRs), firewalls, controllers, load balancers, probes, etc.

6.9. Conclusion

This chapter on virtualization products shows that this new paradigm is inevitable, mainly because of its automation from a central controller. Of course, the security of the whole system must be taken into account. This security requires duplication, but very often triplication of controllers with synchronization between them. These products are very efficient and can greatly improve security and of course the cost.

It should be noted that virtualization products generally consume more energy than their physical counterparts for the same power.

6.10. References

Avramov, L. and Portolani, M. (2014). *The Policy Driven Data Center with ACI: Architecture, Concepts, and Methodology*. Cisco Press, Indianapolis, IN.

Badotra, S. and Panda, S.N. (2020). A survey on software defined wide area network. *International Journal of Applied Science and Engineering*, 17(1), 59–73.

Baktir, A.C., Ozgovde, A., Ersoy, C. (2017). How can edge computing benefit from software-defined networking: A survey, use cases, and future directions. *IEEE Commun. Surv. Tutor.*, 19(4), 2359–2391.

Ellawindy, I. and Heydari, S.S. (2019). QoE-Aware real-time multimedia streaming in SDWANs. In *IEEE Conference on Network Softwarization (NetSoft)*, Paris.

Gibson, D. (2015). *SD-WAN for Dummies*. John Wiley & Sons, Inc., New York.

Kreutz, D., Ramos, F., Verissimo, P., Rothenberg, L., Azodolmolky, S., Uhlig, S. (2015). Software-defined networking: A comprehensive survey. *Proceedings of the IEEE*, 103(1), 14–76.

Lombardi, F. and Di Pietro, D. (2011). Secure virtualization for cloud computing. *Journal of Network and Computer Applications*, 34(4), 1113–1122.

Lowe, S., Marshall, N., Guthrie, F., Liebowitz, M., Atwell, J. (2013). *Mastering VMware vSphere 5.5*. SYBEX, Indianapolis, IN.

Mine, G., Hai, J., Jin, L., Huiying, Z. (2020). A design of SD-WAN-oriented wide area network access. In *International Conference on Computer Communication and Network Security (CCNS)*.

Nagpal, P., Chaudhary, S., Verma, N. (2020). Architecture of software defined wide-area network: A review. *GRD Journals: Global Research and Development Journal for Engineering*, 5(6).

Perez, R., Zabala, A., Banchs, A. (2021). Alviu: An intent-based SD-WAN orchestrator of network slices for enterprise network. In *IEEE 7th International Conference on Network Softwarization (NetSoft)*.

Rajagopalan, S. (2020). An overview of SD-WAN load balancing for WAN connections. In *Fourth International Conference on Electronics, Communication and Aerospace Technology*.

Wen, H., Tiwary, P.K., Le-Ngoc, T. (2013). *Wireless Virtualization*. Springer, Cham.

Woo S., Uppal, S., Pitt, D. (2015). *Software-Defined WAN*. John Wiley & Sons, New York.

Wu, X., Lu, K., Zhu, G. (2018). A survey on software-defined wide area networks. *J. Commun.*, 13(5), 253–258.

Yang, Z., Cui, Y., Li, B., Liu, Y., Xu, Y. (2019). Software-defined wide area network (SD-WAN): Architecture, advances and opportunities. In *28th International Conference on Computer Communication and Networks (ICCCN)*.

7

OpenFlow, P4, Opflex and I2RS

Networks in Edge and Cloud Networking make extensive use of SDN (Software-Defined Networking). SDN is the solution that decouples the data plane from the control plane. The computations and control algorithms are located no longer directly in the network nodes, routers or switches but in a server located somewhere in the network, called a controller. In this chapter, we will study the signaling protocol between the controller and the SDN nodes. The controller can be a physical device or a virtual device located in a data center of varying size.

The data plane includes the forwarding of packets. This solution makes the hardware independent from the software networks and takes into account many parameters. Indeed, this decoupling leads the controller, usually centralized, to retrieve precise information from each client in the network and making complex calculations thanks to load and quality of service algorithms, so that each network flow has its own forwarding tables in the network nodes. Taken to the extreme, each flow could have its own tables that take into account all the characteristics of the client generating that flow: performance, energy consumption, reliability, security, resiliency, cost, etc. This solution does not currently scale, but it could become so with strong decentralization and adapted protocols to allow communication between controllers. We will now detail the OpenFlow protocol.

7.1. OpenFlow signaling

OpenFlow is basic signaling, designed by Nicira, the company behind the SDN technology. This signaling is described in Figure 7.1.

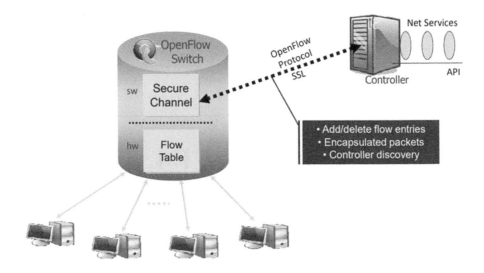

Figure 7.1. *The OpenFlow signaling protocol. For a color version of this figure, see www.iste.co.uk/haddadou/edge.zip*

Figure 7.1 shows an OpenFlow Switch and an OpenFlow Controller using a secure SSL link to allow mutual authentication of the two devices communicating. The term switch is used for the nodes in the SDN network, which are therefore called the OpenFlow Switch.

The OpenFlow Switch has two parts: one that contains the queues, frame senders and frame receivers, with associated flow tables, and one that governs communication with the controller through the OpenFlow signaling protocol. The first part contains all the elements necessary for the physical transport of frames from node to node, and also contains the flow table that allow packet flows to be directed to the correct output queues. There is one line of the flow table for each flow. The OpenFlow protocol provides the necessary elements from the controller to create, manage and destroy the rows of the flow table. This flow table is usually a switching table, but it can be a routing table. The second part concerns the communication between the OpenFlow switch and the controller as well as the information that must pass through it.

As we have already mentioned, the OpenFlow protocol uses a secure SSL/TLS channel to authenticate the two ends of the communication, thus greatly limiting attacks on the communication in progress and imposing mutual authentication. The OpenFlow protocol provides the means to identify packet flows through levels 1, 2, 3 and 4, information that is carried in OpenFlow frames. The signaling also carries actions to indicate what to do with the packets in the flow. OpenFlow also sends very precise statistics back to the controller so that the path determination algorithm can do its job with near-perfect knowledge of the network state. Because the controller is centralized, a small amount of time is required to convey the data.

Figure 7.2 shows a diagram of an OpenFlow network. The controller is connected to all the nodes in the network. Eventually, some nodes may not be OpenFlow enabled and form a subnetwork through which OpenFlow signaling flows. In general, nodes are OpenFlow enabled: they know how to interpret OpenFlow commands and enter actions on flows into flow tables or transform them to fit routing tables or switch tables. Packet flows never pass through the controller unless there is a DPI (Deep Packet Inspection) built into the controller that would analyze all the packets in the flow. In the diagram in Figure 7.2, no flow passes through the controller, except for the first few packets, which are used to determine the requests and the type of flow to be taken into account.

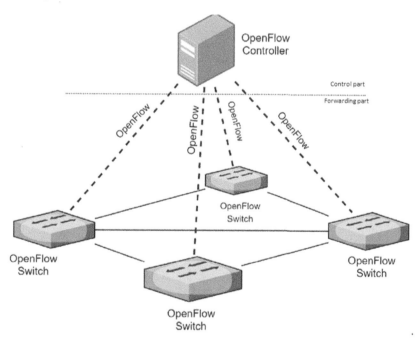

Figure 7.2. *The OpenFlow protocol in a network. For a color version of this figure, see www.iste.co.uk/haddadou/edge.zip*

The fields of the OpenFlow frame are described in Figure 7.3. It shows the flow table which contains six types of information: the match field, the instruction set, the counters, the priority, the maximum timeout and the cookie value. This is for OpenFlow version 1.3 which is the one used today. The three basic forms of information that existed at the beginning, in version 1.0, are given by the first three fields.

The flow is described by the source port number, then the Ethernet addresses, the VLAN number, the VLAN priority, the IP addresses, the protocol and service type fields and finally the TCP or UDP transport layer port numbers. This is defined in the OpenFlow 1.0 protocol. In Figure 7.3, we also see the extensions that were added in version 1.3 that concern the MPLS Label, the experimental part of MPLS and the GRE (Generic Routing Encapsulation) tunnel.

The instruction set has been enriched with each new release. Among the many possible actions, carried by the OpenFlow signaling, the most usual ones include the following:

– send a packet to a list of ports;

– add/reject/modify a VLAN Tag;

– delete a packet;

– send a packet to the controller.

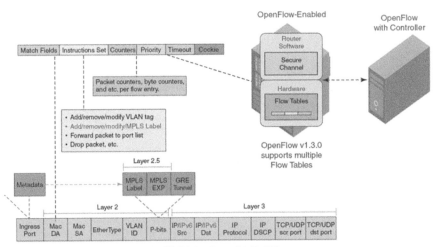

Figure 7.3. *OpenFlow protocol fields 3.0. For a color version of this figure, see www.iste.co.uk/haddadou/edge.zip*

The flow table is completed by frame counters, byte counters, etc., indicating the exact flow statistics on all ports of the network. Thus, the controller has a very accurate and complete view of the entire network it manages.

The first OpenFlow generation (v1.0) was standardized by the Open Network Foundation (ONF) in 2009. Since then, new versions have been released, mainly concerning extensions:

– Version v1.1 in 2011 took into account the MPLS and Ethernet Carrier Grade Q-in-Q techniques that we will see later. Multicast and management are also better considered, as well as the possibility of having multiple routing tables.

– Version v1.2 introduced the ability to control networks using the IPv6 protocol. This version also provided much more detail on flow statistics and introduced multiple controllers where the network could be split into several parts, each of which is handled by a controller.

– Version v1.3 further extended the range of supported protocols such as MAC-in-MAC of the Ethernet Carrier Grade and the possibility to have multiple communication channels between a controller and each of the OpenFlow switches in the network.

– Version 1.4 took a new direction with the control of optical networks.

– Version 1.5 was released in 2015 with some extensions in particular on the treatment of the table of flows and the feedback of statistics.

These different versions of the OpenFlow protocol are summarized in Figure 7.4.

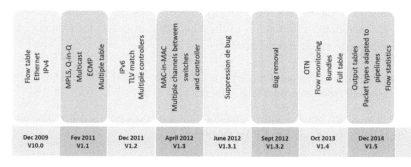

Figure 7.4. *The different versions of OpenFlow. For a color version of this figure, see www.iste.co.uk/haddadou/edge.zip*

One of the difficulties is the processing of the connections, which spans 12 fields in the basic version and 15 fields in the latest version. To speed up the processing of the matching field, the flow table is broken down into 12 or 15 much smaller tables, as shown in Figure 7.5.

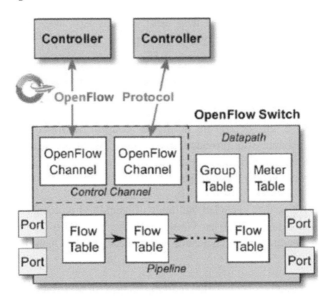

Figure 7.5. *Decomposition of the flow table in OpenFlow. For a color version of this figure, see www.iste.co.uk/haddadou/edge.zip*

In addition to these flow tables, there are group tables which are located between two elementary flow tables as shown in Figure 7.6. These group tables make it possible to add additional information to the flow table in order, for example, to carry out a multipoint or delete certain packets.

Many open-source software programs are natively OpenFlow or OpenFlow compatible. We will mention only a few of them:

– Indigo is an Open Source implementation that runs on a physical machine and uses the characteristics of an ASIC to run OpenFlow;

– LINC is an Open Source implementation that runs on Linux, Solaris, Windows, MacOS and FreeBSD;

– Pantou is OpenFlow for a wireless environment OpenWRT;

– Of13softswitch is Ericsson's soft switch;

– XORPlus is an Open Source Software Switch;

– Open vSwitch is an Open Source software switch developed by Nicira and integrated into the Linux kernel (from version 3.3). Open vSwitch is widely used by many equipment manufacturers in their architecture.

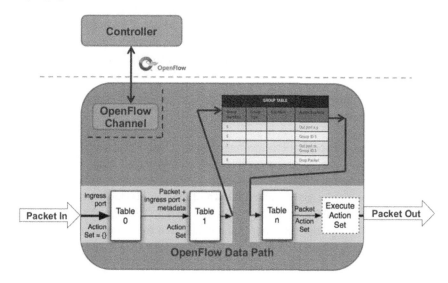

Figure 7.6. *A table of groups in the flow table. For a color version of this figure, see www.iste.co.uk/haddadou/edge.zip*

Also, many controllers are OpenFlow compatible. Among the long list, here are some examples:

– NOX was the first OpenFlow controller.

– FlowVisor is a Java-based OpenFlow controller that acts as a transparent proxy between an OpenFlow switch and multiple OpenFlow controllers.

– POX is a Python-oriented OpenFlow controller with a high-level SDN interface.

– FloodLight is an OpenFlow controller in Java.

– OpenDaylight, which we have already described in the open-source projects, provides a scalable platform that contains a controller at its center. The OpenDaylight platform is used by many companies as a controller, even if they sometimes have other proprietary solutions. This platform can be considered today as the standard, but it remains little used compared to the proprietary controllers of large manufacturers like Cisco or VMware.

– ONOS (Open Network Open System) is one of the latest projects launched as an open-source switch. Its main feature is that it has a distributed operating system that acts as an eastbound-westbound interface. OpenFlow nodes can depend on two controllers, one primary and one secondary. The number of switches supported by the ONOS controller potentially increases with the number of controllers since end-to-end control is achieved through multiple controllers linked together by the distributed operating system.

– Tungsten Fabric is the open-source version of Open Contrail that we saw briefly in Juniper's solution for its SDN.

In addition to the controllers mentioned above, there are many developments underway that are OpenFlow compatible but support many other northbound or southbound interfaces that have been developed to meet the needs of companies. The OpenFlow protocol is certainly the best known of the southbound interface protocols, but nothing is yet finalized in the SDN forward march. This signaling protocol has the advantage of being relatively simple and well suited for its intended use.

7.2. P4

P4 (Programming Protocol-independent Packet Processors) is a specific language in the SDN environment to unambiguously define the data plane configuration regardless of the underlying hardware. This language defines the protocol headers and the processing logic. It enables the automatic reconfiguration of nodes through a compilation providing information and instructions on the actions to be taken.

Although P4 may seem like a general-purpose programming language, it is not; nor is it the successor to OpenFlow that some have named OpenFlow2.0. It can represent any configuration for processing packets in SDN switch nodes. The P4 architecture is shown in Figure 7.7 alongside the OpenFlow structure.

P4 was created in 2013 by the P4 Language Consortium, a nonprofit organization formed by a group of engineers and researchers from Google, Intel, Microsoft Research, Barefoot Networks, Princeton and Stanford. They needed a standard open programming language. The P4 language was used to precisely define how packets are transmitted within the SDN infrastructure. Since then, P4 has been adopted by market leaders and has gained strong support, from cloud providers to telecom operators. Recently, the P4 Language Consortium has been integrated into the Open Networking Foundation (ONF), to carry a consistent SDN set.

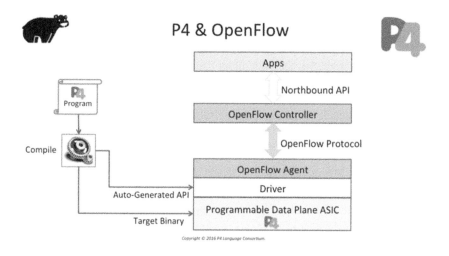

Figure 7.7. *P4 and OpenFlow architectures. For a color version of this figure, see www.iste.co.uk/haddadou/edge.zip*

P4 programmability allows the user to develop new custom features, reduce complexity by removing unnecessary features and tables, and provide better management using diagnostics, telemetry, OAM and more. Modularity allows the user to compose the packet forwarding behavior from libraries, and since the forwarding behavior is specified once, it can be compiled on many devices.

7.3. OpFlex

OpFlex is a completely different signaling protocol for transporting the configuration decided by the controller to the network nodes. This protocol was developed by CISCO and is used as the southbound interface in its SDN architecture, which is called ACI (Application Centric Infrastructure). CISCO then proposed it as a southbound interface standard to the IETF, the protocol specification body for the Internet world.

OpFlex includes an open-source agent that can be integrated into the SDN environment. For example, it is integrated into Open vSwitch (OVS) and OpenDaylight. OpFlex provides a declarative control instead of an imperative control in the other protocols of the southbound interface. A declarative control specifies the policy to be followed but does not indicate the details of how to implement it, unlike an imperative policy, where it is necessary to indicate directly what to do, as is the case with OpenFlow.

OpFlex is therefore the central protocol for configuring Cisco SDN nodes. A business-level policy comes to the APIC controller, which translates the business policy into a network policy. This policy is sent by OpFlex to the nodes, which must translate and complete it to configure the nodes.

7.4. I2RS

I2RS (Interface to the Routing System) is yet another southbound interface, proposed by the IETF to allow a better compatibility of IP networks with SDN technology. The basic idea is to try to reconcile the distributed techniques of the Internet world with the centralized technologies of SDN. From this approach, it is necessary to retain the appearance of an I2RS controller which passes from a distributed mode to a centralized mode as soon as necessary. Indeed, the controller has a complete view of the network to be configured and can detect global problems faster than a distributed system can. These problems can come from both performance and security. For example, the detection of a DDOS (Distributed Denial of Service) is more likely to be taken into account quickly by the controller, which can then take control and impose a configuration adapted to the current attack.

Figure 7.8 describes the architecture of the I2RS environment. The I2RS client is located in the controller, while the I2RS Agent is located in the SDN node.

The data plane uses the Forwarding Information Base (FIB) table to direct frames to the next node. There are two possibilities to fill this FIB: the standard network routing solution like OSPF or IS-IS and the I2RS agent, which receives the flow table from the controller through the I2RS southbound interface. The decision to take one or the other solution to fill the FIB is decided by the controller, which plays the role of a control tower. Globally, as long as the network with its distributed solution gives satisfaction and allows each customer to have the good quality of service and the requested security, there is no modification of the control which remains distributed. As soon as the network is no longer satisfactory, the controller takes over and imposes its flow table.

This solution is the one desired by the IETF to find a compromise between Internet solutions favoring distribution and the centralized solution of SDN with I2RS Clients and Agents available in the controllers and nodes of an SDN network. However, nearly eight years after the standard was defined, very few implementations have been made, and no major industry players are offering it commercially.

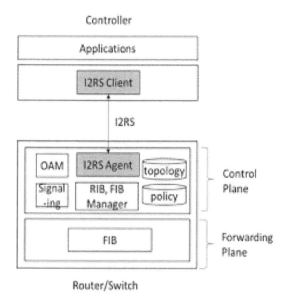

Figure 7.8. *I2RS Architecture. For a color version of this figure, see www.iste.co.uk/haddadou/edge.zip*

7.5. Conclusion

This chapter is devoted to the new southbound interfaces developed over the last 10 years. There are many others, either linked to an equipment manufacturer, or as free software or from de jure or de facto standards. For example, SNMP, which is the management system for IP environments, is an acceptable solution. However, the number of requests required to transport the information needed to configure the nodes would be prohibitive in terms of time, performance and security.

OpenFlow signaling, which has been the main standard since the beginning of SDN, had a rough time when the company Nicira, which had designed the protocol, was acquired by VMware. The other major vendors in the field moved to their own solutions : Juniper to XMPP, Cisco to OpFlex. After several years of divergence on the southbound interface, there has been a return to OpenFlow. Even Cisco adopted OpenFlow in addition to OpenDaylight, for a short year, before abandoning this solution.

It is highly likely that P4 will be the successor to OpenFlow, given the potential of this new solution to keep pace with changes in the software and hardware used in networks.

7.6. References

Bai, J., Zhang, M., Li, G., Liu, C., Xu, M., Hu, H. (2020). FastFE: Accelerating ML-based traffic analysis with programmable switches. *Proceedings of the Workshop on Secure Programmable Network Infrastructure, SPIN'20*. Association for Computing Machinery, New York, USA.

Ben Basat, R., Ramanathan, S., Li, Y., Antichi, G., Yu, M., Mitzenmacher, M. (2020). PINT: Probabilistic in-band network telemetry. *Proceedings of the Annual conference of the ACM Special Interest Group on Data Communication on the Applications, Technologies, Architectures, and Protocols for Computer Communication*, New York, USA.

Bianchi, G., Bonola, M., Capone, A., Cascone, C. (2014). OpenState: Programming platform-independent stateful OpenFlow applications inside the switch. *ACM SIGCOMM Computer Communication Review*, 44(2), 44–51.

Chen, X., Kim, H., Aman, J.M., Chang, W., Lee, M., Rexford, J. (2020). Measuring TCP round-trip time in the data plane. *Proceedings of the Workshop on Secure Programmable Network Infrastructure*, New York, USA.

Ding, D., Savi, M., Siracusa, D. (2019). Estimating logarithmic and exponential functions to track network traffic entropy in P4. *IEEE/IFIP Network Operations and Management Symposium (NOMS)*, Budapest, Hungary.

Ding, D., Savi, M., Antichi, G., Siracusa, D. (2020). An incrementally deployable P4-enabled architecture for network-wide heavy-hitter detection. *IEEE Transactions on Network and Service Management*, 17(1), 75–88.

Hyun, J., Van Tu, N., Hong, J.W.-K. (2018). Towards knowledge-defined networking using in-band network telemetry. *NOMS 2018-2018 IEEE/IFIP Network Operations and Management Symposium*, Seoul, Korea.

Kannan, P.G. and Chan, M.C. (2020). On programmable networking evolution. *CSI Transactions on ICT*, 8(1), 69–76.

Karaagac, A., De Poorter, E., Hoebeke, J. (2019) In-band network telemetry in industrial wireless sensor networks. *IEEE Transactions on Network and Service Management*, 17(1), 517–531.

Kfoury, E.F., Crichigno, J., Bou-Harb, E., Khoury, D., Srivastava, G. (2019). Enabling TCP pacing using programmable data plane switches. *42nd International Conference on Telecommunications and Signal Processing (TSP)*, Dublin, Ireland.

Kim, Y., Suh, D., Pack, S. (2018). Selective in-band network telemetry for overhead reduction. *IEEE 7th International Conference on Cloud Networking (CloudNet)*.

Li, Y., Miao, R., Liu, H.H., Zhuang, Y., Feng, F., Tang, L., Cao, Z., Zhang, M., Kelly, F., Alizadeh, M.Y. (2019). HPCC: High precision congestion control. *Proceedings of the ACM Special Interest Group on Data Communication*.

Marques, J.A., Luizelli, M.C., da Costa Filho, R.I.T., Gaspary, L.P. (2019). An optimization-based approach for efficient network monitoring using in-band network telemetry. *Journal of Internet Services and Applications*, 10(1), 12.

McKeown, N., Anderson, T., Balakrishnan, H., Parulkar, G., Peterson, L., Rexford, J., Shenker, S., Turner, J. (2018). OpenFlow: Enabling innovation in campus networks. *ACM SIGCOMM Computer Communication Review*, 38(2), 69–74.

Niu, B., Kong, J., Tang, S., Li, Y., Zhu, Z. (2019). Visualize your IP-over-optical network in real time: A P4-based flexible multilayer in-band network telemetry (ML-INT) system. *IEEE Access*, 7, 82413–82423.

Open Networking Foundation (n.d.). About [Online]. Available at: https://www.opennetworking.org.

Open Networking Research Center (n.d.). ONRC [Online]. Available at: http://onrc.net.

Pan, T., Song, E., Bian, Z., Lin, X., Peng, X., Zhang, J., Huang, T., Liu, B., Liu, Y. (2019). Int-path: Towards optimal path planning for in-band network-wide telemetry. *IEEE INFOCOM 2019-IEEE Conference on Computer Communications*.

Scholz, D., Oeldemann, A., Geyer, F., Gallenmüller, S., Stubbe, H., Wild, T., Herkersdorf, A., Carle, G. (2020). Cryptographic hashing in P4 data planes. *ACM/IEEE Symposium on Architectures for Networking and Communications Systems (ANCS)*.

Shahzad, S., Jung, E.-S., Chung, J., Kettimuthu, R. (2020). Enhanced explicit congestion notification (EECN) in TCP with p4 programming. *International Conference on Green and Human Information Technology (ICGHIT)*.

Tan, L., Su, W., Zhang, W., Lv, J., Zhang, Z., Miao, J., Liu, X., Li, N. (2021). In-band network telemetry: A survey. *Computer Networks*, 186, 107763.

Tang, L., Huang, Q., Lee, P.P. (2020). Spreadsketch: Toward invertible and network-wide detection of superspreaders. *IEEE INFOCOM, IEEE Conference on Computer Communications*.

Teixeira, R., Harrison, H., Gupta, G., Rexford, J. (2020). PacketScope: Monitoring the packet lifecycle inside a switch. *Proceedings of the Symposium on SDN Research*.

Wang, W., Tammana, P., Chen, A., Ng, T.S.E. (2019). Grasp the root causes in the data plane: Diagnosing latency problems with SpiderMon. *Proceedings of the Symposium on SDN Research*.

Zhang, X., Cui, L., Wei, K., Tso, F.P., Ji, Y., Jia, W. (2020). A survey on stateful data plane in software defined networks. *Computer Networks*, 184, 107597.

8

Edge and Cloud Networking Operators

Telecom operators were quickly interested in virtualization and cloud networking technologies to centralize applications and be able to control and maintain them easily. The specifications for 5G have been particularly influenced by these new technologies. In this chapter, we will examine the impact of Edge and Cloud Networking on 5G technology and, more generally, on telecom operators.

8.1. Edge Networking in 5G architecture

5G architecture is depicted in Figure 8.1. It contains four main parts: the first part is the radio, that is, the transmission from a mobile device to the antenna. 5G radio technology is essentially an enhancement of 4G radio using the improved OFDMA (orthogonal frequency-division multiple access) technique. The throughput has been multiplied by a factor of 10, which allows a very high throughput on the radio but unfortunately slowed down in the non-standalone version, that is, the version that uses the 4G network in the access network and the core network. The high throughput of the radio will be useful as soon as all the components of 5G, that is, the RAN (Radio Access Network) and the 5GC (5G Core network), are available around 2024–2025.

The second part is precisely the access network or RAN that connects the antenna to the core network of 5G. The revolution of 5G is found here. It comes from the virtualization of all processes in an operator data center, called MEC (Multi-access Edge Computing).

The third part is the core network (5GC), which is the network that connects the MEC data centers together to enable communication between 5G antennas and to transport a user's data from an entry point to a potentially very distant exit point.

Finally, the last part is services. This is the first generation of mobile networks where a multitude of services have been specified in the 3GPP release 17 of June 2022. The 3GPP (3rd Generation Partnership Project) is the international organization that specified the 3G, 4G and 5G technologies. These services have the characteristic of being positioned in three main categories: mission critical, Internet of Things and very high-speed mobility. The first category corresponds to applications that require extremely low latency, that is, applications that need a latency time of the order of a millisecond. The latency time is the time between the moment of departure of the packet from the transmitting equipment and the moment of arrival of the same packet in the destination equipment. This interval does not include the time needed to execute the request in the data center if there is a need for intermediate processing between the sender and the receiver. In this class of mission-critical services, we find, for example, controls in the vehicular world or in Industry 4.0 environments. These controls require latencies of less than or equal to one millisecond.

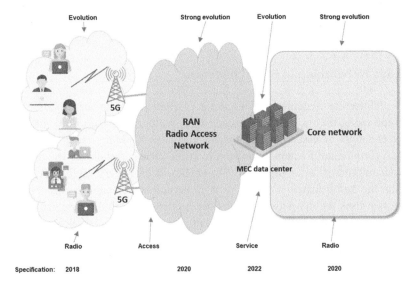

Figure 8.1. *5G architecture. For a color version of this figure, see www.iste.co.uk/haddadou/edge.zip*

The second class of service concerns the Internet of Things with its billions of sensors and actuators to connect. By mid-2022, there were about 37 billion connected objects and by 2025, we should reach a hundred billion connected objects. There is great diversity between objects that can range from a vehicle to a temperature sensor. Similarly, the associated flows range from a few tens of Mbps like an aircraft engine or a vehicle to a few bytes every hour or even every day or month, or even a zero flow for a fire sensor that may not detect any fire during its entire life.

The third class includes all connections and in particular high mobile speeds. Indeed, in the fourth generation, mobile speeds are rather low. The fifth generation will bring a mobile speed, in a high-speed train for example, almost equivalent to that which users could have at home or in their office.

Figure 8.2 characterizes the transition from 4G to 5G. In 4G, there are no data centers at the Edge level. There are only a few data centers in the core of the network, with the primary function of managing and controlling all the network components. 5G brings together a dense set of data centers on the border between the Edge and the core network. These data centers are located about 10 km away from the antenna, so that the latency can be in the millisecond range to support applications belonging to the real-time process category with sub-millisecond latency.

This chain of MEC data centers will be a necessary passage from 2025 onward to allow a user to reach the core network and thus communicate with all the data centers of the major web manufacturers and in particular those of GAFAM. This barrier can be circumvented in different ways such as using satellite constellations, making an alliance with 5G operators or by setting up between the customer and the MEC data center.

This chain of data centers marks the real revolution of 5G by providing many services and by supporting all the processes necessary for the life of a company, from the virtual machines of the digital infrastructure to the application services, including all the management processes, control, automation, intelligence, etc.

We can also see in Figure 8.2 that 5G has a unique technique for connecting optical fibers from both antennas and Wi-Fi access points found in Internet boxes or in enterprise access boxes. This unique technique is the PON (Passive Optical Network) which uses passive optical stars that scatter the light in all directions except the direction it comes from. It is a broadcast network where only the access card that has the correct address can retrieve the information.

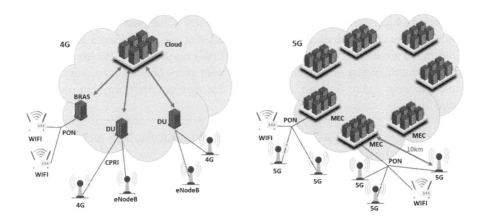

Figure 8.2. *The transition from 4G to 5G. For a color version of this figure, see www.iste.co.uk/haddadou/edge.zip*

Virtualization mainly takes place in the RAN where all intermediate devices between the terminal and the data center disappear. They are replaced by virtual machines that run in the MEC data center. As we have seen in Chapter 1 of this book, there are three types of virtual machines: the virtual machines of the digital infrastructure and in particular managing the signal processing and the level two equipment, that is, at the frame level, the virtual machines of the infrastructure services such as management, control, artificial intelligence, security, etc., and finally the application virtual machines, which include all the new features of 5G, such as vehicle control, remote surgery and the Industrial Internet of Things (IIoT).

8.2. Cloud RAN

Figure 8.3 introduces RAN architecture, which is also called C-RAN or Cloud-RAN. All the functions of RAN are handled by the MEC data center. In particular, signal processing, the functions for routing the signals from the antenna to the data center, then the management and control functions completed by security, intelligence and automation.

It should be noted that all the equipment between the antenna and the data center have disappeared. There is no longer even an electronic cabinet on the antennas to process the signal and manage functions such as geolocation, frame processing, handovers, attachments. Therefore, the signal from the antenna reception has to be transported directly to the data center for processing. As a result, the architecture described in Figure 8.3 is not suitable for existing cabling, which is usually Ethernet based. For this purpose, RoF (Radio over Fiber) is used, which transports the digitized signal directly from the antenna to the data center. If we take a bandwidth of 20 MHz, which is an average value for transporting large flows, a number of samples equal to at least twice the bandwidth, that is, 40 million samples, are required for digitization. If we code a sample on 64 bits, this gives a throughput between the 5G antenna and the data center of 2.560 Gbit/s. We understand the need to use an optical fiber to obtain such data rates, which can be even higher, as we will see in the next chapter.

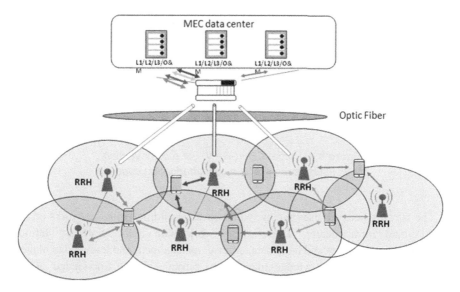

Figure 8.3. *C-RAN architecture. For a color version of this figure, see www.iste.co.uk/haddadou/edge.zip*

The 5G specifiers propose a partial Cloud-RAN as shown in Figure 8.4. In this model, the signal is not transported by the RoF technique but by an EoF (Ethernet over Fiber) technique or directly by an Ethernet technique. In the first case, a digitization of the signal is performed as soon as the signal is received by the antenna and the bits are encapsulated in an Ethernet frame that is sent over the optical channel of the fiber. In the second case, an Ethernet infrastructure is used to transport the IP packets generated in the terminal equipment and encapsulated in an Ethernet frame.

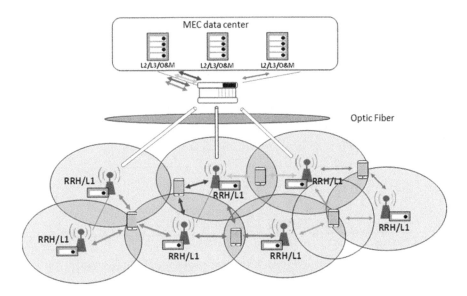

Figure 8.4. *Partial Cloud-RAN. For a color version of this figure, see www.iste.co.uk/haddadou/edge.zip*

The 5G core network uses slicing, which means building virtual networks on top of a digital infrastructure. These slices work with SDN (Software-Defined Networking) technology and have data, control, management and knowledge planes. The objective of the knowledge plane is to retrieve all the knowledge of the network, that is, the information with its context, in order to automate the configurations.

The 5G services are located in the MEC (Multi-access Edge Computing) data center, which connects 5G antennas within a 10-km radius, as well as Wi-Fi access points and even terrestrial fiber optic access. This means that the 5G core network becomes the only network for transporting information, except for Defense networks.

8.3. Cloud Networking at the heart of 5G

The network has a rather revolutionary core for two properties: slicing and the use of Cloud Networking. A slice is a virtual network whose purpose is to support a specific service. The virtualization of the components and functions of the virtual network takes place in MEC data centers. The technology chosen for these virtual networks is SDN, which uses a controller to configure the virtual network nodes.

Figure 8.5 represents the first generation of the 5G core network. It is built on a single slice at first and then quickly on three slices. The main slice is general enough to support the connection of devices that need high throughput whether fixed or mobile.

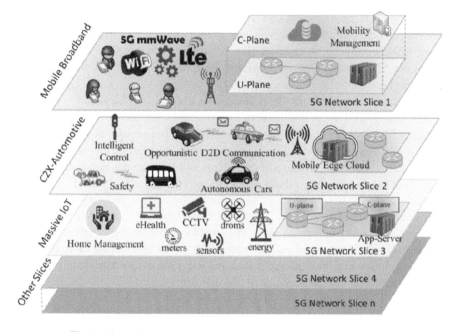

Figure 8.5. *5G slicing with the three main layers. For a color version of this figure, see www.iste.co.uk/haddadou/edge.zip*

The second slice refers to the automation of vehicle driving. Indeed, telecom operators want to take over the automatic driving service of vehicles because behind it there are all the applications of the vehicular world. The third slice focuses on the Internet of Things, with its billions of objects to be connected. In fact, these three basic layers should give rise to many more precise slices in terms of the quality of service offered. For example, in the Internet of Things slice, the throughput of a sensor, such as a thermometer or a pressure analyzer, is totally different from the throughput coming from an object such as an aircraft engine, which constantly transmits an average throughput of around 5 Mbps. It is increasingly necessary to distinguish between RedCap (Reduced Capacity) objects and FulCap (Full Capacity) objects, which require very different qualities of service and will eventually have their own slice. Similarly, in the vehicular world, transmissions related to the braking of one vehicle to the next vehicle have no resemblance to a request for directions at the next traffic light.

Each slice has its own data plan, control plan, management plan and, finally, knowledge plan. One of the foreseeable difficulties of this architecture is its relative complexity since the slices can be very different from each other. The multiplexing of resources between all the slices is also a difficulty that requires further research before being optimized. A third barrier to overcome comes from the size of the slices since SDN cannot control very large networks. It will be necessary to be able to interconnect controllers between them to reach the desired sizes. For this, it is necessary to develop in depth the eastbound-westbound interfaces and in particular the standard coming from the eastbound-westbound interface of the OpenDaylight controller. Finally, the connection of slices from different operators is needed to enable the implementation of virtual networks in all countries of the world. A priori, the operators should respect the SDN standards and have similar slices, but we must wait a little before knowing if these conditions will be met.

6G should amplify the use of virtualization. There are many proposals on the table with still plenty of time to move in a specific direction. However, one solution that seems logical is to build on the momentum of 5G by creating more and more slices associated with multiple applications. These slices are horizontal. To these should be added vertical slices associated with businesses.

6G virtual machines will be placed on the Cloud Continuum by optimizing various application criteria such as performance, energy consumption, security, availability.

8.4. The Cloud and the new Ethernet and Wi-Fi generations

Access to data centers requires speeds that are increasing at a dizzying pace. In the early 2000s, 1 Gbps Ethernet was heavily used, then the speed increased to 10, then 40, then 100 Gbps. The increases continued to reach 400 and then 800 Gbps. The main standards are briefly described in the following paragraphs.

100BaseT4 was created to increase the throughput of 10Mbps or 10BaseT networks using Category 3 (Cat3) cabling to 100 Mbps without having to replace the cabling. Using all four twisted pairs of cabling, two of the four pairs are configured for half-duplex transmission (data can only travel in one direction at a time). The other two pairs are configured as simplex transmission, which means that the data only moves in one direction.

The 100 Mbps type 100BaseTX is also called Fast Ethernet. It transmits data at 100 Mbps. Fast Ethernet works almost identically to 10BaseT, with a physical star topology using a logical bus. 100BaseTX requires Category 5 (Cat5) or higher UTP (Unshielded Twisted Pair) cabling. It uses two pairs of four wires: one to transmit data and one to receive data.

100BaseFX is known as Fast Ethernet over fiber. 100BaseFX requires multimode fibers to carry data. Multimode optical fibers use LEDs to transmit, and they are large enough that the light signals bounce off the edges of the fiber. Bouncing off the edges of the optical core limits the distance between two repeaters.

1 Gbit/s or Gbit1000BaseT or 1GbE is simply called Gigabit Ethernet. It uses a UTP cable of category 5 (Cat5) or higher. It uses all four twisted pairs of cable. The 1GbE uses a physical star topology with a logical bus. There is also 1000BaseF, which uses multimode optical fiber. It supports both duplex and half-duplex data transmission modes.

10 Gbps is known as 10 Gigabit Ethernet or 10GbE. It uses Category 6 (Cat6) or higher UTP cable. It uses the four twisted pairs of UTP cable to provide 10 Gbps speed using only full duplex mode.

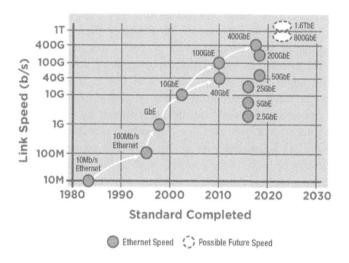

Figure 8.6. *Major Ethernet product lines. For a color version of this figure, see www.iste.co.uk/haddadou/edge.zip*

40 Gbps Ethernet can use single mode optical fiber to reach 40 km but also a metallic cable on 7 m. Finally, 100 Gbps is available up to distances of 100 km on single-mode optical fiber.

400 GbE is divided into two solutions using four or eight channels, that is, 8 or 16 optical fibers: 400Base-DR4 or 400Base-SR8. 800GbE has been available since 2022, and Google is starting to use it on its data center fabrics.

Figure 8.6 groups the major Ethernet product lines.

8.5. Enterprise 5G Edge Networks

Companies are interested in developing their own cloud environment on the Edge, which led to Fog data centers. Initially, it was about taking connections from objects and pre-processing them before sending that information to a core network data center sometimes hundreds or even thousands of miles away. Today, fog computing consists of taking charge of the company's applications directly on its premises because of small data centers located at most a few hundred meters from the users. This solution preserves a strong security of the company's data and leads to extremely fast processing with a totally negligible latency time.

The local area network, based on Ethernet and Wi-Fi, has long been the solution to consolidate the flows to the exit of the company premises. Today, a strong development is taking place around private 5G by its simplicity using the wireless environment and therefore avoiding the deployment of cabling. The private 5G network can be implemented in two ways. The first is to deploy a private 5G network that is physically independent of the telecom operators' 5G networks. In this case, the 5G private network can be implemented by a network or telecom equipment manufacturer or even an Edge Networking provider. For example, AWS markets a private 5G associated with an Amazon data center (AWS private 5G). This data center can be local, in the company such as the Fog data center or located very close to the company in Amazon's premises or in a company allied to Amazon.

The second method is to create 5G private networks using resources from telecom operators. In this case, the telecom operator builds the 5G private network for the company and the telecom operator's MEC data center owns the software resources.

In the first case, the company deploys its entire 5G network with its components on its campus with the use of a frequency located either in the free bands and more particularly in the new 6 GHz band, or in a band reserved for companies by the state.

The realization of the private 5G network is most often taken care of by companies specialized in networks and telecommunications. The advantage of the private 5G solution comes from the ultra-low latency and ultra-connectivity allowing the creation of new business applications or the optimization of existing ones. In particular, Industry 4.0 applications are simply accommodated by 5G. Figure 8.7 shows the two types of 5G: private and operator.

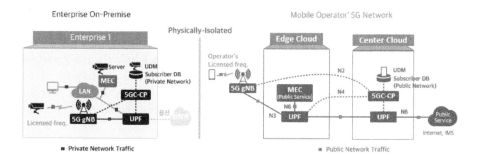

Figure 8.7. *Comparison of public and private 5G. For a color version of this figure, see www.iste.co.uk/haddadou/edge.zip*

Among the advantages of private 5G are privacy and security. The private network is physically separated from the public network and the data stored in the company is safe if the firewall is properly configured. In addition, the use of a data center in the company, linked to the private 5G antenna, allows the company's business applications to be run locally. Even if external communications are cut off, the company can continue to operate normally.

There are also a few drawbacks in private 5G. First, the cost per user, although decreasing, is still higher than the cost of an Ethernet/Wi-Fi network. It is possible that the cost will become competitive with virtualization and the transformation of hardware into software and the arrival of open source like O-RAN. Because of this new software generation, the competition between many companies invested in 5G in software should bring down the costs significantly. It is also necessary to have specialized staff to maintain the 5G environment.

It is possible to have mixed solutions. For example, the company can use an operator's MEC data center while having a private 5G environment in the sense of the antenna but not the support of the company's processes. In this case, the

5G antenna is connected directly to the MEC data center via fiber. The business applications run as virtual machines located in the operator's MEC data center. Another solution is to have a 5G antenna accompanied by a gNodeB, plus a high-speed Ethernet-type connection to the operator's data center.

8.6. Conclusion

Telecom operators are heavily involved in Edge and Cloud Networking. It is clear that it is going to be increasingly difficult to distinguish between Cloud providers and telecom operators. They are entering each other's playgrounds: operators are increasingly integrating the Edge and are even trying to take the place of Cloud providers. Similarly, in the other direction, Cloud providers are moving into the Edge and setting up telecom systems to directly access their own Cloud.

In 2022, this battle is still limited since the tendency of operators is to seek expertise from Cloud providers who inevitably have a better knowledge of this area which is at the center of their interests. In the other direction, Cloud providers are implementing more and more telecommunications systems that are virtualized in their data centers. The 2020s will show us whether a union will have taken place or whether the two approaches will be in conflict.

8.7. References

Aazam, M., Harras, K.A., Zeadally, S. (2019). Fog computing for 5G tactile industrial Internet of Things: QoE-Aware resource allocation model. *IEEE Trans. Ind. Inform.*, 15, 3085–3092.

Agiwal, M., Roy, A., Saxena, N. (2016). Next generation 5G wireless networks: A comprehensive survey. *IEEE Commun. Surveys Tutorials*, 18(3), 1617–1655.

Athonet (2018). SGW-LBO solution for MEC taking services to the edge. White Paper, Athonet.

Cattaneo, G., Giust, F., Meani, C., Munaretto, D., Paglierani, P. (2018). Deploying CPU-intensive applications on MEC in NFV systems: The immersive video use case. *Computers*, 7(4), 558.

Chiang, M. (2017). *Fog for 5G and IoT*. Wiley, New York.

Cominardi, L., Deiss, T., Filippou, M., Sciancalepore, V., Giust, F., Sabella, D. (2020). MEC support for network slicing: Status and limitations from a standardization viewpoint. *IEEE Commun. Standards Mag.*, 4(2), 22–30.

ETSI (2018). Mobile edge computing (MEC); deployment of mobile edge computing in an NFV environment. White Paper, ETSI, Sophia Antipolis, France.

ETSI (2019a). Developing software for multi-access edge computing. White Paper, ETSI, Sophia Antipolis, France.

ETSI (2019b). Multi-access edge computing (MEC): Framework and reference architecture. White Paper, ETSI, Sophia Antipolis, France.

ETSI (2019c). System architecture for the 5G system (5GS). ETSI, Sophia Antipolis, France.

Huazhang, L., Zhonghao, Z., Shuai, G. (2019). 5G Edge Cloud networking and case analysis. In *2019 IEEE 19th International Conference on Communication Technology (ICCT)*, Xi'an, China.

Ksentini, A. and Frangoudis, P.A. (2020). Toward slicing-enabled multiaccess Edge computing in 5G. *IEEE Netw.*, 34(2), 99–105.

Li, S., Da Xu, L., Zhao, S. (2018). 5G Internet of Things: A survey. *J. Ind. Inf. Integr.*, 10, 1–9.

Li, D., Hong, P., Xue, K., Pe, J. (2019). Virtual network function placement and resource optimization in NFV and edge computing enabled networks. *Comput. Networks*, 152, 12–24.

Maglogiannis, I., Iliadis, L., Pimenidis, E. (2020). *An Intelligent Cloud-Based Platform for Effective Monitoring of Patients with Psychotic Disorders*. Springer, Cham, Switzerland, pp. 15–24.

Markakis, E.G. and Mastorakis, G. (2017). *Cloud and Fog Computing in 5G Mobile Networks: Emerging Advances and Applications*. IET Press, Stevenage, UK.

Moura, J. and Hutchison, D. (2019). Game theory for multi-access edge computing: Survey, use cases, and future trends. *IEEE Commun. Surveys Tuts.*, 21(1), 260–288.

Pham, Q.V., Fang, F., Ha, V.N., Piran, M.J., Le, M.B., Hwang, W.J., Ding, Z. (2019). A survey of multi-access edge computing in 5G and beyond: Fundamentals, technology integration, and state-of-the-art. *IEEE Access*, 8, 116974–117017.

Satyanarayanan, M. (2017). The emergence of Edge computing. *Computer*, 50(1), 30–39.

Taleb, T., Samdanis, K., Mada, B., Flinck, H., Dutta, S., Sabella, D. (2017). On Multi-access Edge Computing: A survey of the emerging 5G network Edge Cloud architecture and orchestration. *IEEE Commun. Surveys Tuts.*, 19(3), 1657–1681.

Tomaszewski, L., Kukliński, S., Kołakowski, R. (2020). A new approach to 5G and MEC integration. *Proc. Artif. Intell. Appl. Innov. AIAI IFIP WG Int. Workshops in Advances in Information and Communication Technology*, Hersonissos, Crete, Greece.

Xu, X., Zhang, X., Liu, X., Jiang, J., Qi, L., Bhuiyan, M.Z. (2021). Adaptive computation offloading with edge for 5G-envisioned Internet of connected vehicles. *IEEE Transactions on Intelligent Transportation Systems 2021.*

Zhu, Y., Hu, Y., Schmeink, A. (2019). Delay minimization offloading for interdependent tasks in energy-aware cooperative MEC networks. *IEEE Wireless Communications and Networking Conference (WCNC)*, 15, 1–6.

9

Cloud Networking Protocols

With the advent of digital architectures and Cloud Networking, traditional protocols based on IP packet routing and MPLS (Multiprotocol Label Switching) are being replaced by new generations that either use transport at level one, that is, at the physical layer, or that use Internet protocols but adapt them to have centralized control and a different control plane than the data plane. This new generation is trying to make its way into the new arsenal of operators and cloud providers.

The first solution that is developing in the digital infrastructure is to transport information at the signal level. The terminal sends its frame in the form of digital signals to an antenna which sends the received signals back over an optical fiber to the data center. The signal processing is achieved in the data center. The main technique used to transport the signal is RoF (Radio over Fiber). Another solution that can be seen as intermediate between existing and RoF is to keep signal processing at the antenna and to transmit the information in binary form (0 and 1) in an Ethernet frame. The electronic cabinet that carries out the signal processing emits Ethernet frames filled with binary data without understanding the transported data. We are dealing with a level 2 transport (frame level). The advantage of this last technique is using the existing networks in Ethernet form without having to rewire the environment in an optical fiber.

The new transport protocols are primarily used inside data centers. They must enable very high-speed connections between servers, which number in the thousands, even tens and hundreds of thousands. The first protocol used by large industrial companies was TRILL (Transparent Interconnection of Lots of Links). It is an Ethernet network with extensions to make routing possible in a level-two Ethernet environment. TRILL can be defined very roughly as an Ethernet network encapsulated in an Ethernet network, which allows for an address hierarchy and routing on the internal network. We will come back to this later.

Cloud and Edge Networking,
by Kamel HADDADOU and Guy PUJOLLE. © ISTE Ltd 2023.

Then, there are the inter-data center networks, which also have huge throughputs and can go from one subnet to another. There are a number of possibilities that appear in the literature and in the reality of cloud managers. In general, these are extensions of VLAN (Virtual LAN) techniques to be able to keep VLAN numbers consistent, even when the data centers are not in the same subnets. In addition, the number of VLANs is limited by the numbering zone, which only allows 4,096 VLANs using only 12 bits of this zone. In this category of protocols, we find the following:

– VXLAN (Ethernet in UDP) (User Datagram Protocol) initially proposed by VMware;

– NVGRE (Ethernet in IP) at the initiative of Microsoft;

– 11aq (EVLAN) from the Ethernet Carrier Grade standardization.

To this list, we must add at least the following two protocols:

– MPLS from operators and standardized by both the ITU and the IETF, possibly with the SDN option for its openness;

– LISP (IP in IP) supported by many OEMs and operators.

The last protocol LISP (Locator/Identifier Separation Protocol) is fundamentally different from the previous ones because it is a level-three protocol that is an encapsulation of an IP packet in an IP packet. This encapsulation is explained by the need to have two IP addresses associated with an end machine. The inside address is for identification of the recipient and the outside address is for routing. The latter address designates the physical location of the receiver. This solution makes it possible to move virtual machines while keeping the same identity and changing the IP address of the virtual machine's location. We will also come back to this with more details in the following. Finally, the upstream protocol corresponds to the one developed for SDN: it is OpenFlow, which was introduced by the company Nicira, which was later acquired by VMware.

We will examine in the following the protocols of levels 1 then 2 (Ethernet level) and finally 3 (IP level).

9.1. Low-level protocols

The low-level protocols essentially contain the transport of signals directly on the optical fiber or through a frame but always directly on the optical fiber.

9.1.1. *Radio over Fiber*

RoF is a protocol used heavily with virtualization to allow a radio signal to be sent to a data center for signal processing. In particular, 5G antennas are connected to MEC data centers over fiber with the use of RoF. Figure 9.1 shows an example of connecting an antenna on the left to send a radio signal to the equipment that will process the signal.

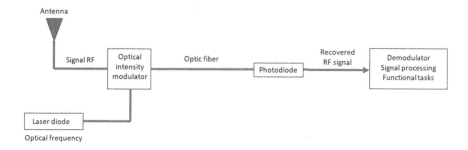

Figure 9.1. *The RoF technique. For a color version of this figure, see www.iste.co.uk/haddadou/edge.zip*

As we mentioned in the previous chapter dealing with the Edge and Cloud and 5G, the signal to be transmitted can be extremely complex since it is the superposition of all the signals that arrive at the antenna with the attenuations and interferences that occur on the signals. The only way to send it to the data center without losing quality is to digitize it. To do this, we must use the sampling theorem which tells us that we must sample at least twice the bandwidth. For a bandwidth of 100 MHz, we need 200 million samples. These samples must then be coded, for example, on 64 bits, which gives a data rate of 12.8 Gbps. This data rate can be highly compressed with high-end hardware located in the antenna. If we want to go very fast in order not to waste time, compression exists but it is much lower in quantity than what we could do with a lot of power. In fact, we have to transport between 100 Mbps and 1 Gbps.

If we want to virtualize a Wi-Fi antenna in a Fog data center, we need to connect the Wi-Fi antenna to the data center via fiber. Since the bandwidth can go up to 320 MHz for a Wi-Fi 7, we need to sample at 640 million times per second, and once compressed, the data rates are more than 1 Gbps.

It should be noted that the antennas are not passive and that there are actually calculations to be made to digitize the signal received by the antenna. A second solution is to use Ethernet to connect the antenna to the data center. In this case, the signal processing is done in the antenna as in 4G. In this case, electronic cabinets are required at the foot of the antenna.

9.1.2. *Ethernet over Fiber*

EoF (Ethernet over Fiber) is the solution used when Ethernet networks are already in place without the enterprise or between the antennas and the Fog data center in general. Figure 9.2 describes the EoF technique. It consists of connecting antennas to a data center via an optical fiber. In general, a star topology is used, but it does not optimize the length of fiber used. The optimization can be done by a loop in which the Wi-Fi access points are connected on one wavelength of the optical fiber.

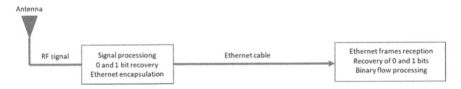

Figure 9.2. *The EoF technique. For a color version of this figure, see www.iste.co.uk/haddadou/edge.zip*

9.2. Virtual extensible LAN

We will now look at a number of protocols for the Cloud and more specifically for data center interconnection. VLANs are often used but with extensions to avoid the strong limit of the number of VLANs supported on a network, which is limited by the length of the VId (VLAN identifier) field (12 bits), that is, 4,096 VLANs maximum (in fact, 4,094 because the values 0 and 4,095 are reserved for signaling). To have much larger networks, the VId zone must be extended to double its length or even quadruple it. One of the most used extensions concerns VXLAN.

VXLAN (Virtual eXtensible LAN) is a solution to realize communications in Clouds between data centers. The technology is quite similar to that of VLANs and Carrier Grade Ethernet extensions. It was initially proposed by Cisco and VMware. The disadvantage of the basic IEEE 802.1Q VLAN standard solution is the limitation to 4,094 networks. The VXLAN solution allows the basic technology to be extended in parallel with Carrier Grade Ethernet.

In the VXLAN solution, to go from one subnet to another subnet we want to be able to cross any network, and to do this we must go back to the transport level. This is why the Ethernet frame to be transported between data centers is encapsulated in a UDP message which is itself encapsulated in an IP packet and then in an Ethernet frame. These successive encapsulations are described in Figure 9.3.

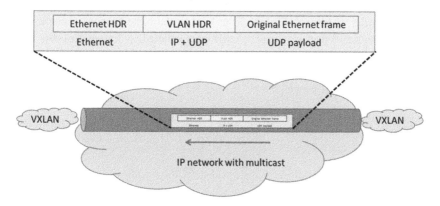

Figure 9.3. *The VXLAN protocol. For a color version of this figure, see www.iste.co.uk/haddadou/edge.zip*

To the basic Ethernet frame, a 24-bit VXLAN identification zone is added, allowing more than 16 million VLANs to be reached, followed by a 64-bit UDP field, which is then encapsulated in an Ethernet frame with fields, source address, destination and VLAN number, associated with VXLAN. The basic Ethernet frame thus forms the "data" part of the UDP message, which is itself carried in an IP packet encapsulated in an Ethernet frame. This property makes it possible to create VLANs beyond an Ethernet domain. These encapsulations are shown in Figure 9.2. The overhead is significant, as we see, since the VXLAN frame adds 36 bytes to the base frame.

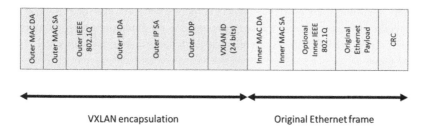

Figure 9.4. *The VXLAN encapsulation. For a color version of this figure, see www.iste.co.uk/haddadou/edge.zip*

9.3. Network Virtualization using Generic Routing Encapsulation

Another protocol also supported by the IETF concerns NVGRE (Network Virtualization using Generic Routing Encapsulation). It is supported by several manufacturers, including Microsoft. Like the previous one, this protocol enables crossing an intermediate network between two data centers, this network being an IP network. In order to keep the value of the VLAN, the basic Ethernet frame must be encapsulated in an IP packet, which is itself encapsulated in frames to cross the IP network. The original Ethernet frame ends up unchanged in the VLAN on the other side of the network. Figure 9.5 describes the tunnel that is opened in the IP network to transport the original frame.

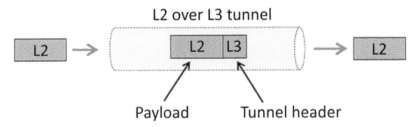

Figure 9.5. *The NVGRE protocol. For a color version of this figure, see www.iste.co.uk/haddadou/edge.zip*

Classic solutions such as Ethernet MEF (Metro Ethernet Forum), GVLANs (generalized VLANs) and Ethernet Carrier Grade are all suitable for interconnecting data centers. Let us start with the Ethernet MEF solution.

9.4. Ethernet MEF

MEF (Metropolitan Ethernet Forum) networks have been around for a long time. Originally, their purpose was to interconnect corporate networks in a metropolis at very high speeds. However, they are also very suitable for interconnecting data centers with each other. MEF Ethernet networks use switched Ethernet networks at 1, 10, 40 and 100 Gbps.

To implement IP telecommunications and services requiring strong time constraints, Ethernet Carrier Grade networks enable the introduction of priorities, with the difference that the IEEE 802.1p field, which is used to introduce these priorities, has only 3 bits. These 3 bits only allow eight priority levels, compared to the 14 levels defined by the IETF for DiffServ services.

The priority levels proposed by MEF are as follows:

– 802.1p-6 DiffServ Expedited Forwarding;

– 802.1p-5/4/3 DiffServ Assured Forwarding;

– 802.1p-5, which has the lowest loss;

– 802.1p-3, which has the highest loss;

– 802.1p-2 DiffServ Best Effort.

In the Ethernet environment, flow control is generally a difficult problem. Various proposals have been made to improve it. Back-pressure methods propose to send control messages from overloaded switches, allowing the related switches to stop sending to the congested node for a time specified in the control primitive.

The choice made by the MEF is a frame relay type control, where we find exactly the same parameters:

– CIR (Committed Information Rate);

– CBS (Committed Burst Size);

– PIR (Peak Information Rate);

– MBS (Maximum Burst Size).

This solution shows the excitement of the Ethernet world, which presents a large number of solutions capable of competing with each other to satisfy the new generation needed in the world of cloudified networks.

9.5. Ethernet Carrier Grade

Ethernet was designed for computer applications, not for applications in the world of telecommunications, which requires special qualities called Carrier Grade. In order to meet the demands of operators, the Ethernet environment had to be transformed. We speak of Carrier Grade Ethernet, which refers to a solution that is acceptable to telecommunications operators with the control and management tools required in this case. This transformation mainly concerns switched Ethernet. Carrier Grade Ethernet must have features found in telecommunications networks, including the following:

– Reliability, which allows for very few failures. The mean time between failures, or MTBF (Mean Time Between Failure), must be at least 50,000 h.

– Availability, which must reach the classic values of telecommunications, that is, being in working order 99.999% of the time. This value is far from being reached by traditional Ethernet networks, which are in operation 99.9% of the time.

– Protection and recovery. When a failure occurs, the system must be able to restart after a maximum time of 50 ms. This time is derived from the telecommunications industry, which only accepts outages over time intervals that are shorter than this value. SONET networks, for example, reach this value of reconfiguration time. The solutions generally use redundancy, total or partial, which makes it possible to start another path, foreseen in advance, in the event of cut.

– Performance optimization through active or passive monitoring. Not all performances are homogeneous when packet flows vary. It is therefore necessary to adapt the flows so that they can transit without problems.

– The network must be able to accept SLAs (Service Level Agreements). The SLA is a typical notion in an operator's network when a customer wants to negotiate a service guarantee. The SLA is determined by a technical part, the SLS (Service Level Specification), and an administrative part in which the penalties are negotiated if the system does not give satisfaction.

– Management is also an important feature of operator networks. In particular, fault detection and signaling systems must be available to keep the network up and running.

From a technical point of view, Ethernet Carrier Grade is an extension of VLAN technology. A VLAN is a local network in which machines can be located at very distant points. The goal is to make this network work as if all the points were geographically close to each other to form a local network. A VLAN can contain several users. Each Ethernet frame is broadcast to all the machines in the VLAN. The tables that perform the switching of frames are fixed and can be seen as switching tables in which the addresses of the recipients are references.

When the VLAN has only two points, sending an Ethernet frame from one point to the other is similar to switching along a path. This is the vision that has been adopted in Carrier Grade Ethernet. Paths are formed by determining VLANs. The path is unique and simple if the VLAN has only two points. The VLAN introduces a multipoint if there are more than two points.

The problem with this solution is the limited size of the VLAN zone, which is defined by only 12 bits. This is fine for an enterprise network with standard Ethernet switching, but this size becomes quite insufficient for Carrier Grade Ethernet, which is aimed at operator networks. Therefore, the size of the VLAN field had to be increased.

Ethernet Carrier Grade can be subdivided into several VLAN extension solutions, all of which are described in Figure 9.6. The most traditional solution is to use the IEEE 802.1ad standard, which is known by several names: Ethernet PB (Provider Bridge), QiQ (Q in Q) or cascaded VLAN. The IEEE 802.1ah standard is also known as MiM (MAC-in-MAC) or PBB (Provider Backbone Bridge). The most advanced solution is called PBT (Provider Backbone Transport), or PW over PBT (PseudoWire over Provider Backbone Transport). It enables returning to a classical transport solution in which Ethernet frames are switched according to a succession of references corresponding to MPLS-SL (Service Label). The solutions described in the previous lines are illustrated in Figure 9.6.

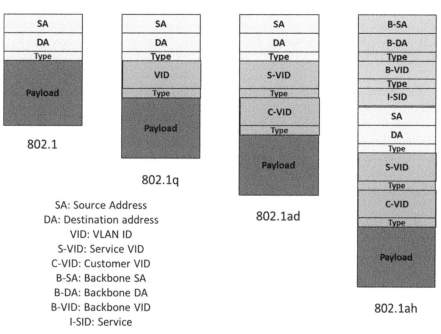

Figure 9.6. *The different versions of Ethernet Carrier Grade. For a color version of this figure, see www.iste.co.uk/haddadou/edge.zip*

We will first consider Ethernet PB (Provider Bride) technology. The ISP adds a VLAN number to that of the customer. There are therefore two VLAN numbers: the C-VID (Customer-VLAN ID) and the S-VID (Service-VLAN ID). The introduction of the service provider makes it possible to extend the notion of VLAN to the operator's network without destroying the user's VLAN. This solution makes it possible to define the broadcasts to be made in the operator's network. Since the field in the

Ethernet frame, in which the VLAN number is indicated, is 12 bits long, this allows up to 4,094 entities in the operator's network to be defined. These entities can be services, tunnels or broadcast domains. However, if 4,094 is a sufficient value in an enterprise, it remains far below the needs of an operator. Some implementations use a reference translation to enlarge the domain, which increases the complexity of managing the whole. This solution is therefore not suitable for large networks.

The solution proposed by the IEEE 802.1ah PBB group improves on the previous one by switching frame traffic to the MAC address. This solution, known as MIM (MAC-in-MAC), encapsulates the client's MAC address in the operator's MAC address. This allows the core operator to know only its MAC addresses. In the PBB network, the correspondence between the user MAC address and the network MAC address is only known by the edge nodes, avoiding the explosion of MAC addresses.

The third solution, called PBT (Provider Backbone Transport), is quite similar to the MPLS technique, while providing the properties necessary for Carrier Grade, such as an unavailability rate of less than 50 ms. It is a kind of MPLS-backed tunnel. The PBT tunnel is created as an MPLS tunnel, with the credentials corresponding to the ends of the network. The client and server VLAN numbers are encapsulated in the MPLS tunnel, which itself may have a carrier VLAN differentiation. The real reference is therefore 24 bits + 48 bits, or 72 bits.

The last solution is the MPLS PseudoWire (PS) service. In this case, the user and operator VLANs are encapsulated in an MPLS service tunnel, which in turn can be encapsulated in an MPLS transport tunnel. This solution comes from the encapsulation of tunnels in MPLS.

Carrier Grade Ethernet technology is of interest to many operators. Time will tell which solution will win. But it is already certain that the encapsulation of VLANs in VLANs will be present in all of them.

9.6. Transparent Interconnection of Lots of Links

TRILL (Transparent Interconnection of Lots of Links) is an IETF standard implemented by nodes called RBridges (bridge-routers) or TRILL switches. TRILL combines the advantages of bridges and routers. In effect, TRILL follows Layer 2 routing using link state. RBridges are compatible with the Layer 2 bridges defined in the IEEE 802.1 standard and may gradually replace them. RBridges are also compatible with IPv4 and IPv6 routers and therefore fully compatible with current IP routers. The routing performed with the IS-IS protocol at level 2 between the

RBridges replaces the Spanning Tree Protocol (STP) in TRILL. Figure 9.7 describes the processes used by the TRILL protocol.

Using Layer 2 routing between Ethernet addresses eliminates the need to configure Layer 3 and associated IP addresses. The link state protocol used for routing can include additional Type, Length, Value (TLV) information. To avoid potential loops during rerouting, RBridges manage a hop count, which is the number of nodes traversed. When this number reaches a maximum value, determined by the operator, the frame is destroyed.

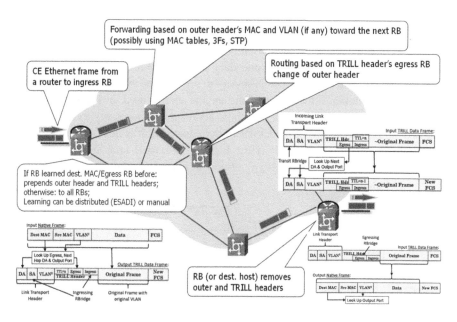

Figure 9.7. *The TRILL protocol. For a color version of this figure, see www.iste.co.uk/haddadou/edge.zip*

When an Ethernet frame arrives in the first RBridge of the network, called IRB (Ingress RBridge), an additional header is added, the TRILL header. The frame will be decapsulated by the output RBridge, or ERB (Egress RBridge). The new frame carries in the TRILL part the address of the ERB, which gives this technology the routing status since it uses the address of the recipient and not a reference. This RBridge address is 2 bytes long, replacing the 6 bytes of the classic Ethernet frame. The addressing is thus achieved on addresses that the user defines themself. The added header consists of 6 bytes in total, ending with the input and output addresses, preceded by the hop count and a flag. The Ethernet frames are recovered after decapsulation at the output. We then find a classic transfer mode, either on the VLAN number, the Ethernet address being used as reference, or by decapsulating the Ethernet frame to find the IP packet.

An interesting point is the possibility of transferring frames of the same flow by several paths simultaneously: it is called multipath, or ECMP (Equal Cost MultiPath), which makes it possible to detect the various routes of the same cost and to direct the frames on these various routes.

9.7. Locator/Identifier Separation Protocol

LISP (Locator/Identifier Separation Protocol) was developed to enable the transport of virtual machines from one data center to another without the virtual machine changing its IP address. It is necessary to separate the two interpretations of the IP address: the identifier of the user machine and the address used for routing (locator). If we want to keep the same virtual machine address, it is necessary to differentiate these two values. This is what the LISP protocol does, but it is not the only protocol to make this separation: the HIP (Host Identity Protocol) and SHIM6 (Level 3 Multihoming Shim Protocol for IPv6) protocols also do it, but with mechanisms based on end machines.

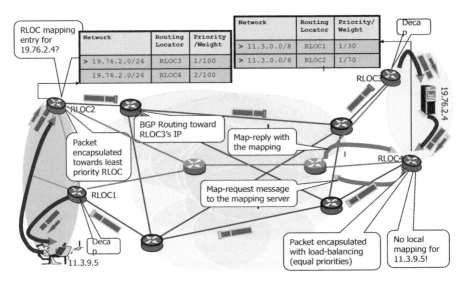

Figure 9.8. *The LISP protocol. For a color version of this figure, see www.iste.co.uk/haddadou/edge.zip*

In the LISP approach, routers support the association between the endpoint address, the EID (Endpoint ID), and the routing address, the RLOC (Routing-Locator). At the start of the communication between an endpoint and a server machine, the traffic is intercepted by an ingress router, the iTR (ingress Tunnel Router), which must determine the address to be used for routing in order to access the server machine whose address is that of the EID. The exit of the network to reach the EID is done by the eTR (egress Tunnel Router). To perform this operation, the router addresses a directory service, the EID-RLOC. Once the location address of the RLOC recipient is obtained, the packet can be routed to this recipient. This process is not visible to the machine and server. Figure 9.8 describes the main elements of the LISP protocol.

Note that LISP allows the use of other addresses than IPv4 or IPv6 level 3, such as a GPS location or a MAC address.

9.8. Conclusion

The networks found in Cloud architectures are essentially Ethernet based and generally use a level-two architecture. In other words, we use less and less the IP protocol, which is much heavier and much slower, with the exception of the LISP protocol, which is defended by many equipment manufacturers and operators because of its good adaptation to the Cloud environment.

With the considerable success of Level 2, the IEEE 802 working group is increasing its offensives in other directions such as wireless networks. Ethernet and Wi-Fi have become the two main standards for access and home networks. In Chapter 11, we will see the new Wi-Fi standards that are arriving and that make it increasingly possible to do without cables, because of both the capacities that exceed Gbps and the mesh structure that allows Wi-Fi access points to be linked directly.

9.9. References

Alberro, L., Castro, A., Grampin, E. (2022). Experimentation environments for data center routing protocols: A comprehensive review. *Future Internet*, 14(1), 29–37.

Bourguiba, M., Haddadou, K., Pujolle, G. (2016). Packet aggregation-based network I/O virtualization for cloud computing. *Computer Communications*, 35(3), 309–319.

Handley, M., Raiciu, C., Agache, A., Voinescu, A., Moore, A.W., Antichi, G., Wójcik, M. (2017). Re-architecting datacenter networks and stacks for low latency and high performance. *Proceedings of the Conference of the ACM Special Interest Group on Data Communication*, Los Angeles, CA.

Hooda, S.K. and Kapadia, S. (2014). *Using TRILL, FabricPath, and VXLAN: Designing Massively Scalable Data Centers (MSDC) with Overlays*. Cisco Press, Indianapolis, IN.

Lucero, M., Parnizari, A., Alberro, L., Castro, A., Grampín, E. (2021). Routing in fat trees: A protocol analyzer for debugging and experimentation. *Proceedings of the 2021 IFIP/IEEE International Symposium on Integrated Network Management (IM)*, Bordeaux.

Mattos, D., Duarte, O., Pujolle, G. (2018). A lightweight protocol for consistent policy update on software-defined networking with multiple controllers. *Journal of Network and Computer Applications*, 122, 77–87.

Medhi, D. and Ramasamy, K. (2017). *Network Routing: Algorithms, Protocols, and Architectures*, 2nd edition. Morgan Kaufmann Publishers Inc., San Francisco, CA.

Meza, J., Xu, T., Veeraraghavan, K., Mutlu, O. (2018). A large scale study of data center network reliability. *Proceedings of the Internet Measurement Conference 2018*, Boston, MA.

Mir, N.F. (2014). *Computer and Communication Networks*. Prentice Hall, Hokoben, NJ.

Pujolle, G. (2013). Metamorphic networks. *Journal of Computing Science and Engineering*, 7(3), 198–203.

Robertazzi, T.G. (2017). *Introduction to Computer Networking*. Springer, Cham.

Sajassi, A., Drake, J., Bitar, N., Shekhar, R., Uttaro, J., Henderickx, W. (2018). A network virtualization overlay solution using ethernet VPN (EVPN); RFC 8365. Document, IETF, Fremont, CA.

Secci, S., Pujolle, G., Nguyen, T.M.T., Nguyen, S. (2014). Performance-cost trade-off strategic evaluation of multipath TCP communications. *IEEE Transactions on Network and Service Management*, 11(2), 250–263.

Singh, H. (2017). *Implementing Cisco Networking Solutions*. Packt Publishing, Birmingham.

10

Edge and Cloud Networking in the IoT

Telecommunications operators are particularly interested in the Internet of Things (IoT), which represents a spectacular market with 37 billion objects connected by mid-2022 and around 100 billion expected by 2025. The first step toward the Cloud was marked by Cisco in 2012 by their introduction of Fog computing. The idea was to locally collect streams from objects and process them in part before sending them to Clouds to run fairly complex applications using large numbers of streams. Fog computing continues to exist through companies that use objects to control applications such as industrial manufacturing automation called Industry 4.0 and more specifically IIoT (Industrial Internet of Things) for all the sensors and actuators needed in industrial manufacturing automation.

Telecommunications operators have started to position themselves in the object connection market with 4G and LTE-M (machine). LTE-M derived from the 4G standard continues to be marketed by operators especially for complex objects that need a high throughput. The last generation of 4G saw the birth of NB-IoT (Narrow Band Internet of Things) to connect objects, but the processing of streams from these objects was pushed to data centers far from the connected objects, simply because Edge technologies were not present at the operators. At the same time, proprietary solutions with LoRa and SigFox were already on the market to connect objects that only transmit low data rates. Extensions of these two solutions to include the processing of flows in small data centers enabled moving toward fog computing.

With 5G and the arrival of MEC data centers, the outlook changes completely since operators can treat the processes managed by the sensors in very short time frames since the MEC data centers are located at a distance less than 10 km from the object. The latency times remain in the order of a millisecond, which truly controls manufacturing robots or various machines with very strong time constraints.

The objective of 5G is to take the Internet of Things market, including Industry 4.0, the smart city, meter readings that number in the billions, the control of door opening in vehicles or homes, and so on. The preponderance of telecom operators is clearly indicated in the long-term forecasts for everything that is connecting RedCap (Reduced Capability) and FulCap (Full Capability) objects. Before we look at the use of Edge and Cloud Networking in the IoT, let us look at the networks that will be used.

10.1. Internet of Things networks

We can examine the different categories of networks for the IoT in Figure 10.1. In this figure, we discover the networks of private long-distance operators with proprietary solutions like LoRa or SigFox. The latter company went bankrupt in 2022, which shows that it is not easy to impose oneself in this emerging field. We also have Wi-Fi technology, which is very present in the IoT, with several categories of networks, including the product dedicated to the IoT, which bears the commercial name of HaLow.

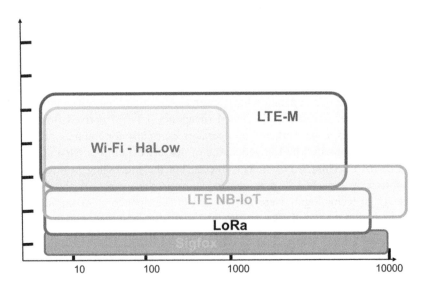

Figure 10.1. *Network categories in the Internet of Things. For a color version of this figure, see www.iste.co.uk/haddadou/edge.zip*

This HaLow network is more specifically found in enterprises to connect all objects of the company on a Fog data center with a maximum range of 1 km. In addition to the standard frequencies of the free 863 kHz band, Europe has added frequencies in the 923 MHz band that was already available in the United States. Finally, we have the two telecom operator networks LTE-M and NB-IoT. As these are networks that carry up to about 10 km, they are linked to MEC (Multi-access Edge Computing) type data centers. The IoT flow management processes are virtual machines or containers located in these data centers.

If we look at the forecasts, the IoT networks that are favored by prospective studies are mainly NB-IoT with a strong increase in market share and LoRa. However, in the medium term, the proportion of LoRa networks should decrease and be replaced by HaLow, which relies on the popularity of Wi-Fi and its very low connection cost. HaLow enables a range of about 1 km and the connection of a few thousand objects.

Before going into a little more detail about the categories of networks for the IoT, let us look through Table 10.1 at the different requirements for the major services that are expected from the IoT.

	Energy	Range	Cost	Throughput	Location	Type of protocole	Security
Health	Years	Body Area Network	Flexible	Weak	Strong location	Simple	Critical
Home Building	Months	Tens of meters	Faible	Weak	Weak	IP possible	Good
Surveillance	Months	Tens of meters	Moyen	Strong	Non in general	IP	Good
Industrial control	Days	Building size	Élevé	Medium	Non in general	IP	Good
Marketing	Days	Tens of meters	Moyen	Weak	Precise	IP Possible	Weak
Energy management	Months	Local and long distance	Moyen	Weak	Weak	Simple	Weak
Remote monitoring	Months	Long	Faible	Weak	Rather weak	IP	Very strong security

Table 10.1. *The different characteristics of Internet of Things networks. For a color version of this figure, see www.iste.co.uk/haddadou/edge.zip*

In this figure, the first line is about health. We can see that the energy supply must be provided by a battery capable of lasting for years. The range corresponds to a Body Area Network (BAN), for example, the network of sensors on user bodies. The cost can be quite high since health is a high priority, the throughput does not need to be high and is often counted in a few tens or hundreds of bit/s, the localization requires knowing quite precisely where the individuals are, the type of protocol can be very simple and security is critical. The second line is concerned with the home and building and the characteristics are as follows: the energy must be sufficient to hold for periods of time of the order of a year, the range is a few tens of meters, the cost must be relatively low, the throughput does not need to be high, the localization is also not very important. Other areas of interest in this table are surveillance, industrial control, marketing, energy management and remote monitoring.

10.2. Low Power Wide Area Networks

LPWANs (Low Power Wide Area Networks) are a generation of networks for the IoT that started in the early 2010s and aim to connect objects over long distances with very low energy consumption. These are generally low-intelligence and low-cost objects that only use a low data rate. We can cite a few examples to illustrate this category of networks: the management of spaces in a large parking lot, the reading of water or electricity meters, the management of crops in intelligent fields, the management of a fleet of vehicles, etc.

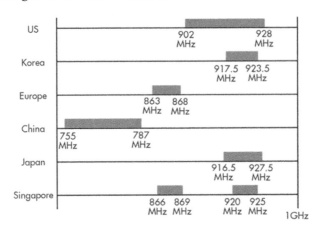

Figure 10.2. *Frequencies used for LPWANs. For a color version of this figure, see www.iste.co.uk/haddadou/edge.zip*

The range of these LPWANs is about 10 km, but obviously depends on the environment, since most of them use the free bands slightly below 1 GHz. These bands are explained in Figure 10.2. They are more or less wide depending on the country, and particular attention must be paid to avoid using products that are not compatible in the country where they are to be deployed.

The cells are large since the frequency is low, but there is an important difference between networks deployed in rural and urban areas : while the range can reach 10 km in rural areas, it can be limited to 1 km in urban areas.

The quantity of messages to be transmitted is very low and corresponds to a few messages per day of limited length. For example, SigFox technology, which was one of the first to try to settle on the marketing and processing from these small objects, allows up to 140 messages per object per day, the load of each message being 12 bytes. The wireless network throughput remains very limited, often at values below 1 kbps.

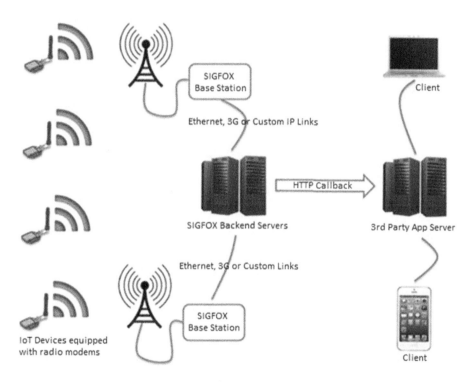

Figure 10.3. *The SigFox solution. For a color version of this figure, see www.iste.co.uk/haddadou/edge.zip*

The SigFox solution is shown in Figure 10.3. In addition to an antenna network using a proprietary solution, there is a set of servers to process the streams coming from the objects. These servers represent the equivalent of a Fog data center. The pre-processed data can then be transmitted to Cloud data centers to perform much more complex services using big data or the search for particularly complex correlations given the huge volumes.

The SigFox solution is intended for objects that are not very intelligent and at the beginning only transmitted to the antenna without a return path. The cost of processors having fallen sharply, today's objects are more intelligent than they were 10 years ago, and the need for a return path to send commands to the object becomes almost mandatory. This is what made the success of the second solution we describe, LoRa, which enables having higher data rates and a full duplex mode at the base with in addition an increased security because of encryption methods.

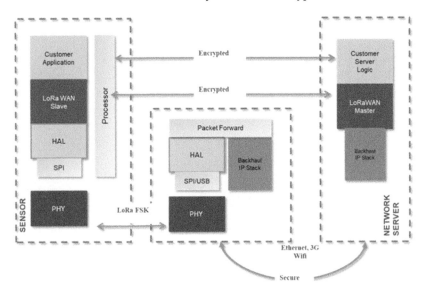

Figure 10.4. *The LoRa solution. For a color version of this figure, see www.iste.co.uk/haddadou/edge.zip*

LoRa technology is described in Figure 10.4. It is also a proprietary technology that has made its way because it has been adopted both by telecommunication operators who had no other solution to offer in the 2010s and by private organizations that have deployed it heavily. We can see in Figure 10.4 that there are two parts: the first is the radio part that goes from the object to the antenna and the second goes from the antenna to a Cloud to realize the processing of the streams coming from the objects.

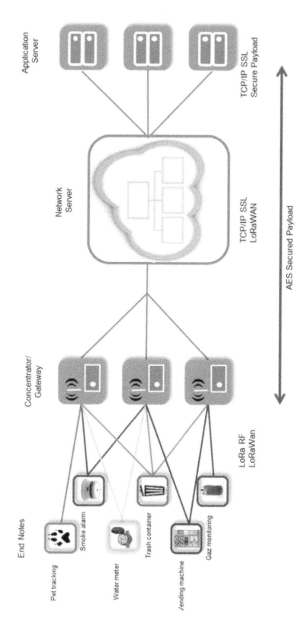

Figure 10.5. *LoRaWAN. For a color version of this figure, see www.iste.co.uk/haddadou/edge.zip*

LoRaWAN is the communication protocol using an intermediate gateway, between the proprietary radio part and the communication with the data center. The security of the communication is ensured at two levels: the first in the LoRaWAN protocol and the second at the application level. The object must be powerful enough to perform the associated encryption. The difference with SigFox comes from both the higher throughput, full duplex and the more important security elements from smarter objects.

The complete LoRaWAN system is described in Figure 10.5. The left-hand side shows the LoRa radio connection, which is a proprietary technology that was originally proposed by the French start-up Cycléo, which was acquired by the company Semtech, which, by setting up the LoRa Alliance, has succeeded in convincing many companies and operators to choose this solution.

One could also classify operator networks as LPWANs, but they are quite different because they are specified completely by 3GPP and are integrated directly into operator technologies. So, we prefer to describe them in a specific section in this same chapter. Before that, we will look at the more classical networks coming from personal and local networks.

10.3. PAN and LAN networks for the IoT

PANs (Personal Area Network) are networks whose range is very low, around 10 m, but can reach 100 m in good conditions in a direct view environment or, on the contrary, be limited to 2 or 3 m. These networks have not been designed to connect directly to data centers to process the throughput coming from the objects but rather to small servers or even to a smartphone, a laptop or a tablet.

We will start with the Bluetooth network, which almost became the standard for PANs after an initial specification by the IEEE 802.15 working group that deals with the standardization of this category of networks. In fact, in 2005, the Bluetooth 1.2 standard was published by the IEEE, but immediately afterward it was stated that this technology could not remain the standard because its throughput was too low. This throughput as we will see is very low: the sum of the throughput of a Bluetooth network is below 1 Mbps.

Figure 10.6 describes the two topologies that a Bluetooth network can take. The first, the piconet, groups together a maximum of eight devices, including a master and seven secondaries. The access technique is polling; the master is the first machine that switches on. Then, as soon as a second machine is turned on, it searches if there is a master around and if there is, it becomes a secondary machine.

There can be a third then a fourth machine that connects, up to a seventh. The master interrogates the secondaries in turn, and if they have something to transmit, they ask the master to give them slots to transmit their packets, more precisely one, three or five slots. Then, the secondary hands over to the master who asks the next secondary. If a secondary has nothing to transmit, it returns to the master.

The advantage of this technique is to avoid collisions between several transmissions to the same antenna and then to allow some synchronization since the secondaries are interrogated in turn, always in the same order. The telephone speech is therefore quite simple to transmit via Bluetooth.

The second topology represented in Figure 10.6 is a scatternet which is in fact an interconnection of piconets. In the overlay network, the masters become secondaries except for one that remains the master of the overlay network. Thus, there is a master, secondary-masters and secondaries.

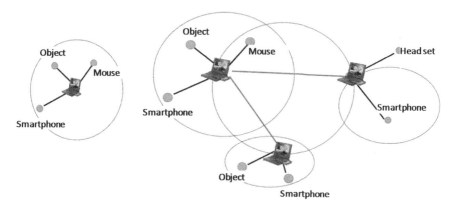

Figure 10.6. *Bluetooth: piconet and scatternet. For a color version of this figure, see www.iste.co.uk/haddadou/edge.zip*

We can also note the use of a bandwidth of 1 MHz per Bluetooth network, the free 2.4 GHz band being cut into 79 bands of 1 MHz. The transmission uses frequency hopping, that is to say, jumping on a new band of 1 Mhz every 625 μs. The advantage is a complex carrier sense, unlike Wi-Fi which uses a fixed band.

Since the abandonment by the IEEE of the standard, Bluetooth is doing well under the aegis of a few major manufacturers such as Ericsson and Apple. In 2022, version 5.3 was the norm. The successive versions are as follows:

– Bluetooth 1.x: less than 1 Mbps (version ratified by the IEEE 802.15 in 2005 before being abandoned);

– Bluetooth 2.0 + EDR, then 2.1 +EDR (2007) (Enhanced Data Rate): 3 Mbps;

– Bluetooth 3.0 + HS (2009) (high speed) = Bluetooth for signaling, Wi-Fi for transmission;

– Bluetooth 4.0 + LE (2010) (low energy) = Bluetooth 2.1+ HS +ULP (ultra-low power);

– Bluetooth 5.0 (BLE + increased data rate).

Another standard, which is also quite successful but growing slowly, comes from ZigBee networks. This technology is certainly among those that consume the least energy possible by optimizing all radio and software elements. In particular, a protocol stack has been developed to replace TCP/IP which consumes a lot of energy because at the time of its conception, there was no question of optimizing energy. The throughput is also very low and the objective is not to transmit large volumes but essentially to transmit control information. ZigBee is found in many controls for sensors and for control devices. Figure 10.7 describes some of the uses of ZigBee.

This technology is rather adapted for the data centers of the embedded Edge. It enables connecting many sensors in a vehicle, for example.

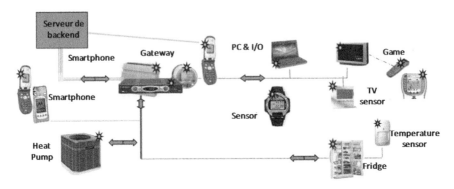

Figure 10.7. *ZigBee. For a color version of this figure, see www.iste.co.uk/haddadou/edge.zip*

The rising technology comes from the Wi-Fi world, with the product whose commercial name is HaLow. There is no precise meaning in this acronym, which however corresponds to Long Range – Low Power. Standardization took place in the 2010s with a first standard in 2017 under the name IEEE 802.11ah. It is a long-range Wi-Fi essentially because of the use of the free band of 863 MHz which reaches 1 km of range in good condition. Figure 10.8 shows the relative ranges of the different Wi-Fi solutions.

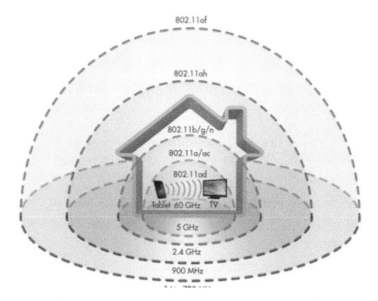

Figure 10.8. *The HaLow range (IEEE 802.11ad). For a color version of this figure, see www.iste.co.uk/haddadou/edge.zip*

The throughput of HaLow is a few hundred kbps per machine to allow several thousand objects to be connected to the central antenna. To avoid too many simultaneous collisions, the CSMA/CD technique is distributed over different zones. In other words, two clients who do not belong to the same zone cannot collide since they transmit on different time slots.

This HaLow technology is clearly aimed at the IoT, and many applications are starting to be implemented since the early 2020s: Smart Grid applications, meter reading, connection of low-capacity objects, sensors, etc. Streams from these objects are stored and processed in data centers that belong to the Fog or Embedded Edge.

An example of the use of HaLow technology is described in Figure 10.9.

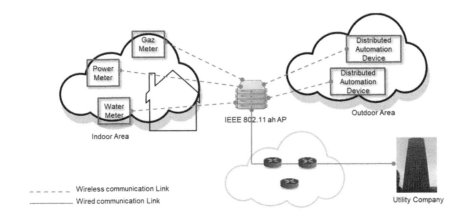

Figure 10.9. *The HaLow environment. For a color version of this figure, see www.iste.co.uk/haddadou/edge.zip*

HaLow Wi-Fi has a bright future due to its origin in the IEEE 802.11 world, its long range and the number of Wi-Fi enabled objects that is increasing rapidly.

10.4. Telecommunications operator networks for the IoT

Operators were very late in getting started on the IoT before integrating solutions into existing technologies. The first 3GPP specifications date back to the 4G release 13. Three main directions are described as follows:

– The eMTC, which is also called LTE-M, corresponds to a relatively simple use of 4G technology by limiting the speed somewhat to 1 Mbps. This solution is mainly used to connect rather powerful objects with high data rates.

– The NB-IoT, the main solution to connect objects, has been thought in this sense. The basic channel has a rate of 250 kbps and this solution is well adapted to all objects with the possibility of going down to extremely low rates and, as we will describe later with solutions of appointment, realizing strong reductions of the energy consumption.

– The EC-GSM-IoT, which uses the channels of GSM or 2G Edge technology to replace telephone circuits with packets at very low data rates, usually around 8 kbps but can go up to 240 kbps.

Figure 10.10 compares these three standards with the two main proprietary solutions.

	sigfox	LoRa	eMTC Rel. 13	NB-IoT Rel. 13	EC-GSM-IoT Rel. 13
Carrier Frequency	Unlicensed ISM: < 1GHz	Unlicensed ISM: < 1GHz	Licensed LTE bands	Licensed LTE/ GSM bands	Licensed GSM bands
Node/Device Bandwidth	100/600 Hz	125/250/500 kHz	1.08 MHz	180 kHz	200 kHz
Max. Data Rate	600 bps	50 kbps	1 Mbps	250 kbps	240 kbps
Range	Many Km	Several Km	Several Km	Several+ Km	Several+ Km
Standardization/Driver	Proprietary / Sigfox	Proprietary / Semtech	3GPP Huawei, Ericsson, Qualcomm, Mediatek, u-Blox, Quectel,...	3GPP Huawei, Ericsson, Qualcomm, Mediatek, U-blox, Quectel,...	3GPP Huawei, Ericsson, GCT, Qualcomm, Quectel,...
Use Cases	Massive devices, small & infrequent data	Massive devices, moderate data rates	Critical (e.g., surveillance Cameras for public safety)	Massive devices, secure LTE commun.	Massive devices, secure GSM commun.

Figure 10.10. *Comparison of different solutions for the Internet of Things. For a color version of this figure, see www.iste.co.uk/haddadou/edge.zip*

These specifications, dating back to 4G, have been improved in the 5G releases, and the real novelty is the introduction of MEC data centers whose objective is to process the streams coming from objects. Where Fog was supposed to be the basic solution, 5G operators would like to regain a dominant position by handling object-related processes in their own data centers. IoT represents a primary focus for operators along with vehicular connections, Industry 4.0 and various mission-critical type applications. In fact, the dual mission-critical plus IoT orientation is often found, for example, in IIoT.

Figure 10.11 describes the key benefits of the NB-IoT solution.

These advantages come from a very low cost per connection, which should go down for about 10 years to reach the value of one dollar per month, a very low energy expenditure allowing an object to use the same battery for 5 to 10 years, an excellent penetration in buildings and a large range coming from the low frequencies used, a simple deployment within the framework of 5G operators and finally a very large number of connections on the same antenna that can exceed 100,000 simultaneously.

The overall throughput can become dizzying and the processing requires large data centers. We will examine the platforms to be implemented in these centers later in this chapter.

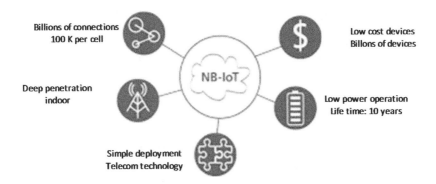

Figure 10.11. *The NB-IoT environment. For a color version of this figure, see www.iste.co.uk/haddadou/edge.zip*

Figure 10.12 describes the technology used in 5G NB-IoT to enable very low energy consumption through a rendez-vous technique that allows the object to turn itself off and back on when the next signal presents itself. If the object is on standby for a long period of time, it will not be accessible from the network during that time. NB-IoT objects are designed for applications where it is acceptable for them to remain inaccessible for long periods of time.

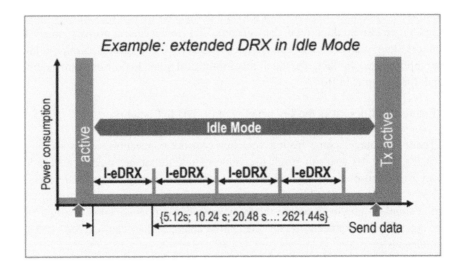

Figure 10.12. *The eDRX technology used in NB-IoT for rendez-vous. For a color version of this figure, see www.iste.co.uk/haddadou/edge.zip*

The eDRX (Extended DRX) solution offers a feature that was introduced by 3GPP in the release 13. It is an enhancement to DRX that was introduced to reduce the power consumption of IoT devices, allowing the object to be in sleep mode for a significant amount of time.

In NB-IoT, eDRX allows the object to sleep for a long time and wake up periodically to check if a communication starts. Since the sleep time in eDRX is considerably longer than in DRX, there can therefore be large energy savings compared to the first generation of rendezvous. The appointment timer in DRX can last up to 2.56 s; the eDRX in connected mode allows up to 9.22 s.

10.5. Platform for the IoT

The network part we just discussed is important, but a processing platform is needed to handle the throughput from the objects and be able to derive applications from them. Figure 10.13 describes the structure of a platform in the general framework. Starting at the bottom of this figure, we obviously discover the communication layer with the objects but also the formatting. Indeed, there is no real standard format applied to the data coming from the objects and it is generally necessary to have a gateway to modify the data format, to allow interconnection between many objects and a coherent presentation of the data at the level of the platform. The next layer takes care of the management of the objects and the necessary updates. The third layer takes care of the data processing and the actions to be performed for the application. Then, it is necessary to be able to follow visually and thus to have a layer for the visualization interface. The layer above concerns the data processing tools that can come from Big Data analytics or various tools from artificial intelligence. Finally, the top layer contains the tools for the external interface with the users of the platform, with SDK (Software Development Kit) and gateways to command systems or other platforms.

The properties of these different layers are summarized as follows.

– Connectivity and standardization: it transforms different protocols and data formats into a single software interface ensuring accurate data transmission and interaction with all end devices.

– Device management: it ensures that connected objects are working properly, continuously executes patches and updates for software and applications running on end devices or edge gateways.

– The database: massive storage of data from connected objects and devices. It guarantees the data requirements in terms of volume, variety, and speed.

– Action execution and management: makes sense of the data with event-driven, rule-based action triggers that allow intelligent actions to be executed based on the sensor database.

– Analytics: performs complex analysis from data collection using machine learning to perform prediction to extract maximum values corresponding to the data streams coming from the IoT.

– Visualization: allows us to see patterns and observe trends from visualization dashboards where data are represented by line, stacked or pie charts, 2D or even 3D.

– Additional tools: enable IoT developers to prototype, test and commercialize IoT solutions by creating applications and platforms to visualize, manage and control connected end machines.

– Outsourcing: integrates with third-party systems and the rest of the IT ecosystem via APIs, SDKs and gateways.

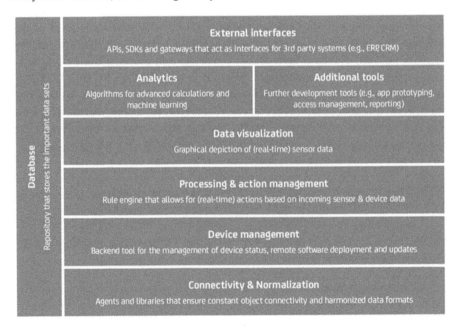

Figure 10.13. *Platform architecture for the IoT. For a color version of this figure, see www.iste.co.uk/haddadou/edge.zip*

The virtual machines that make processing intelligent are briefly described in Figure 10.14. This is mainly Big Data analytics to process huge volumes from many connected objects.

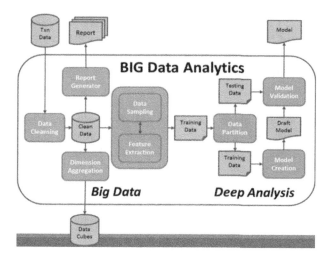

Figure 10.14. *Using AI for the IoT. For a color version of this figure, see www.iste.co.uk/haddadou/edge.zip*

AI control is described in Figure 10.15. It takes the classic form of a stream from a sensor that retrieves data from the IoT environment to send it to a data center located more or less far away depending on the real-time and processing capacity needed. Various mechanisms from artificial intelligence are available as virtual machines. The processing enables obtaining the command that is sent to the actuator, which then exerts its action on the object.

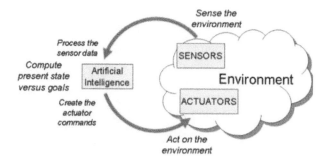

Figure 10.15. *AI control in the IoT. For a color version of this figure, see www.iste.co.uk/haddadou/edge.zip*

Many large web companies have developed their own platforms, especially the big three companies, AWS, Microsoft and Google who share 67% of all processing in the world's data centers.

The first one we will describe is Microsoft's Azure platform. The Azure Internet of Things is a set of Microsoft-managed Cloud services that connect, monitor and control billions of IoT objects. Microsoft's IoT solution is to enable objects to communicate with one or more back-end services hosted in the Azure Cloud.

We need certified objects that can work with Azure IoT Hub. For prototyping, hardware such as an MXChip IoT DevKit or a Raspberry Pi can be used. The Devkit has built-in sensors for temperature, pressure, humidity and a gyroscope, accelerometer and magnetometer. In the same way, the Raspberry Pi allows us to connect many types of sensors.

Microsoft provides open-source device SDKs that can be used to create applications that run on objects. These SDKs simplify and accelerate the development of IoT solutions.

Typically, IoT devices send measurements made by sensors to back-end services in a cloud. However, other types of communication are possible, such as a back-end service sending commands to sensors. In the following, we give some examples of sensor-to-cloud and cloud-to-sensor communication:

– A refrigerated truck sends the temperature every 5 min to a server.

– The monitoring service sends a command to a sensor to change the frequency at which it sends measurements to help diagnose a problem.

– A sensor sends alerts based on the measured values. For example, a sensor monitoring a reactor in a chemical plant sends an alert when the temperature exceeds a certain value.

– The sensors send information to be displayed on a dashboard for viewing by human operators. For example, a control room in a refinery can display temperature, pressure and flow volumes in each pipe, allowing operators to monitor the facility.

Sensor SDKs from IoT support common communication protocols such as HTTP, MQTT, and AMQP. IoT sensors have very different characteristics compared to other environments such as browsers and mobile applications. SDKs help address the challenges of securely and reliably connecting sensors to back-end services. Specifically, IoT sensors are often embedded systems with no human operator (unlike a phone). They can be deployed in remote locations, where physical access is expensive. They can only be reached via the back-end.

An SDK may not get the necessary power or have sufficient processing resources. It may have intermittent, slow or expensive network connectivity. It may need to use proprietary, custom or industry-specific application protocols. So, in an IoT solution, the back-end service must provide features such as:

– receiving large-scale measurements from specialized sensors indicating how to process and store the data;

– analyzing measurements to provide information, if possible in real time;

– sending commands from the cloud to a specific sensor;

– providing resources in the sensors and allowing the sensors to connect directly to the infrastructure;

– *controlling* the status of the sensors and monitoring their activities;

– managing the firmware installed on the sensors.

For example, in a remote monitoring solution for an oil pumping station, the back-end uses measurements from the pumps to detect abnormal behavior. When the back-end identifies an anomaly, it can automatically send a command back to the sensor to take corrective action. This process generates an automated feedback loop between the sensor and the cloud that greatly increases the efficiency of the solution.

The Azure platform is described in Figure 10.16.

Google's platform is quite similar to the Azure platform we just introduced. Google's core product is Cloud IoT Core which is one of the services provided by Google in the IoT space. It is useful when thousands of sensors need to be connected with a necessity to have a scalable and manageable service. One of the best features of Cloud IoT Core is automatic load balancing and data scaling. The two main components of Cloud IoT Core are as follows:

– Device manager: helps register sensors with services and also provides the mechanisms to authenticate sensors. This manager also maintains a logical configuration for each device, and it can also be used to control sensors remotely from the cloud.

– Protocol Bridge: a way for sensors to access or connect to Google Cloud using some standard protocols like HTTP and MQTT. With the help of this equipment, it is possible to use existing sensors with minimal firmware changes.

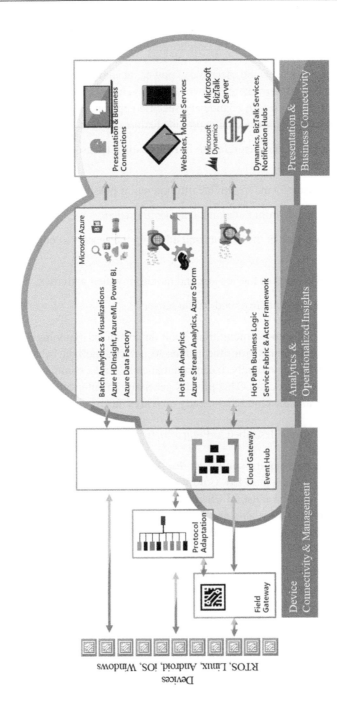

Figure 10.16. *The Azure platform for IOT (Microsoft). For a color version of this figure, see www.iste.co.uk/haddadou/edge.zip*

Google's platform is described in Figure 10.17.

Figure 10.17. *Google's platform for IoT (Google). For a color version of this figure, see www.iste.co.uk/haddadou/edge.zip*

An important feature of Google Cloud Platform comes from functions that are well suited to handling large amounts of data from the IoT and integrate artificial intelligence. Moreover, when creating a project, scalability is the most complex part. Because of its serverless architecture, GCP easily meets the programming requirements of a project. Some of the components of the Google Cloud Platform IoT solution are explained here.

– Grossery:

This is a cloud-based IaaS model designed by Google, used to store and process huge masses of data using multiple SQL (Structured Query Language) queries. In particular, BigQuery is a type of database that is different from transactional databases like MySQL.

– Cloud IoT Core:

This is a fully managed service to easily connect, manage and ingest data from internet-connected sensors. Apart from that, it also enables other services in the Google Cloud platform to collect, process, manage and visualize IoT data in real time.

– Pub/Sub:

This is a type of asynchronous service for communication to other services, used in serverless and microservice-based architectures. In this type of model, any service published on a topic is immediately received by all subscribers to that particular topic. It can be used to implement any event-driven architecture.

– Hardware:

Google offers a version of Android for IoT sensors that is known as Android Things. Android Things helps developers configure and push software updates to the operating system.

– Cloud Functions:

To route data using Pub/Sub, a function-based system is used. A function is nothing more than a java script code triggered when an event takes place. The code runs without the need for an infrastructure manager because everything is handled by Google itself.

AWS also has a platform for managing objects. It is very similar to the previous two, and we will only present it in Figure 10.18.

We will introduce a last platform, the one from IBM. Its objective is similar to the previous ones, that is, to manage the flows coming from sensors and any other objects. It is called the Watson IoT Platform. It gives rise to a fully managed service hosted in the cloud, with the aim of simplifying the valuation of connected objects belonging to the IoT. These objects are automatically connected to the platform, whether it is a sensor, a gateway or any other equipment. With this platform, the collected data are securely sent to the cloud using the Message Queuing Telemetry Transport (MQTT) messaging protocol. The manager can configure and manage the objects using a dashboard or secure APIs so that applications can access and use the data directly. Figure 10.19 provides a diagram of this platform.

Edge and Cloud Networking in the IoT 177

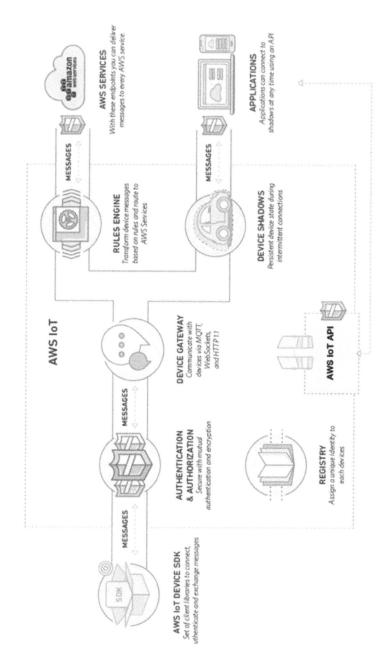

Figure 10.18. *The AWS platform for IOT. For a color version of this figure, see www.iste.co.uk/haddadou/edge.zip*

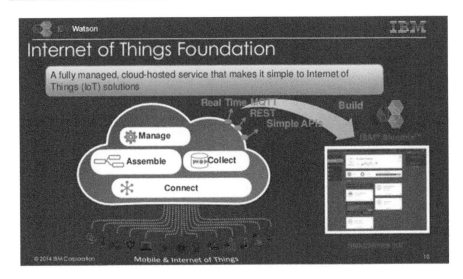

Figure 10.19. *The IBM platform for IoT. For a color version of this figure, see www.iste.co.uk/haddadou/edge.zip*

10.6. Conclusion

The IoT has enabled Fog data centers to take off by collecting data streams from connected objects and processing them to send only the data that need further processing such as Big Data or artificial intelligence to the cloud. Today, the data streams coming from the objects are processed in the Cloud Continuum on the data center that is best able to handle the constraints of the application. While data recovery across many network categories is still far from being mature, the processing part in data centers is now fully mastered.

10.7. References

Anusuya, R., Renuka, D.K., Kumar, L.A. (2021). Review on challenges of secure data analytics in edge computing. *International Conference on Computer Communication and Informatics (ICCCI)*, New York.

Bonomi, F., Milito, R., Zhu, J., Addepalli, S. (2014). Fog computing and its role in the internet of things. In *Proceedings of the First Edition of the MCC Workshop on Mobile Cloud Computing*. Association for Computing Machinery, New York.

Cao, K., Liu, Y., Meng, G., Sun, Q. (2020). An overview on edge computing research. *IEEE Access*, 8, 85714–85728.

Dolui, K. and Datta, S.K. (2017). Comparison of edge computing implementations: Fog computing, cloudlet and mobile edge computing. In *2017 Global Internet of Things Summit (GIoTS)*. IEEE Proceedings, New York.

Goudarzi, M., Wu, H., Palaniswami, M., Buyya, R. (2020). An application placement technique for concurrent IoT applications in edge and fog computing environments. *IEEE Transactions on Mobile Computing*, 20(4), 1298–1311.

Hassan, N., Gillani, S., Ahmed, E., Yaqoob, I., Imran, M. (2018). The role of Edge computing in Internet of Things. *IEEE Commun. Mag.*, 99, 1–6.

Jasenka, D., Francisco, C., Admela, J., Masip-Bruin, X. (2019). A survey of communication protocols for internet of things and related challenges of fog and cloud computing integration. *ACM Computing Surveys (CSUR)*, 51(6), 1–29.

Khan, W.Z., Ahmed, E., Hakak, S., Yaqoob, I., Ahmed, A. (2019). Edge computing: A survey. *Future Generation Computer Systems*, 97, 219–235.

Li, Y. and Wang, S. (2018). An energy-aware edge server placement algorithm in mobile edge computing. In *2018 IEEE International Conference on Edge Computing (EDGE)*, San Francisco, CA.

Li, C., Wang, Y., Tang, H., Zhang, Y., Xin, Y., Luo, Y. (2019). Flexible replica placement for enhancing the availability in edge computing environments. *Computer Communications*, 146, 1–14.

Lin, J., Yu, W., Zhang, N., Yang, X., Zhang, H., Zhao, W. (2017). A survey on internet of things: Architecture, enabling technologies, security and privacy, and applications. *IEEE Internet of-Things (IoT) Journal*, 4, 1125–1142.

Madisetti, V. and Bahga, A. (2014). *Internet of Things: A Hands-on-Approach*. VPT Books, Blacksburg, VA.

Mishra, D., Dharminder, D., Yadav, P., Rao, Y.S., Vijayakumar, P., Kumar, N. (2020). A provably secure dynamic ID-based authenticated key agreement framework for mobile edge computing without a trusted party. *Journal of Information Security and Applications*, 55, 102648.

Mondal, S., Das, G., Wong, E. (2019). Cost-optimal cloudlet placement frameworks over fiber-wireless access networks for low-latency applications. *Journal of Network and Computer Applications*, 138, 27–38.

Porambage, P., Okwuibe, J., Liyanage, M., Ylianttila, M., Taleb, T. (2018). Survey on multi-access edge computing for Internet of Things realization. *IEEE Commun. Surveys Tuts.*, 20(4), 2961–2991.

Qi, B., Kang, L., Banerjee, S. (2017). A vehicle-based edge computing platform for transit and human mobility analytics. *ACM/IEEE Symposium on Edge Computing*, 2017, 1–14.

Rafique, W., Qi, L., Yaqoob, I., Imran, M., Rasool, R., Dou, W. (2020). Complementing IoT services through software defined networking and edge computing: A comprehensive survey. *IEEE Communications Surveys & Tutorials*, 22(3), 1761–1804.

Sha, K., Yang, T.A., Wei, W., Davari, S. (2020). A survey of edge computing-based designs for IoT security. *Digital Communications and Networks*, 6(2), 195–202.

Sun, X. and Ansari, N. (2016). Edge IoT: Mobile edge computing for the internet of things. *IEEE Communications Magazine*, 54(12), 22–29.

Taleb, T., Dutta, S., Ksentini, A., Iqbal, M., Flinck, H. (2017). Mobile edge computing potential in making cities smarter. *IEEE Commun. Mag.*, 55(3), 38–43.

Zhao, X., Shi, Y., Chen, S. (2020). MAESP: Mobility aware edge service placement in mobile edge networks. *Computer Networks*, 182, 107435.

11

Cloud Continuum in Vehicular Networks

Vehicular networks are a rapidly developing environment. The transition to reality is close and will become a reality with the autonomy of vehicles that will need to interconnect to talk to each other and make decisions together. In fact, three types of vehicular networks will be superimposed: the control network that allows the automation of driving, the passenger data network knowing that the driver will become a passenger himself and finally the manufacturer's network whose objective is to follow the lifecycle of the vehicle, to carry out maintenance and to make repairs, sometimes while driving, following a breakdown diagnosis.

The debate to choose the components of these vehicular networks is great because the market is huge. Numerous solutions are being developed, sometimes complementary and sometimes competing. Globally, we can say that three solutions are confronting each other for the vehicle interconnection part. The first is the use of networks connecting vehicles directly without using a large antenna. This solution is illustrated in Figure 11.1.

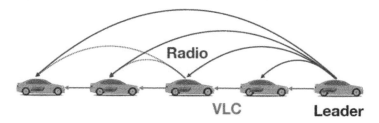

Figure 11.1. *Communication between vehicles: V2V. For a color version of this figure, see www.iste.co.uk/haddadou/edge.zip*

Cloud and Edge Networking,
by Kamel HADDADOU and Guy PUJOLLE. © ISTE Ltd 2023.

The second solution comes from 5G because of its mission critical property which has the possibility of having latencies of the order of 1 ms. This solution requires data to pass through the antenna to access a server. Figure 11.2 illustrates this second possibility.

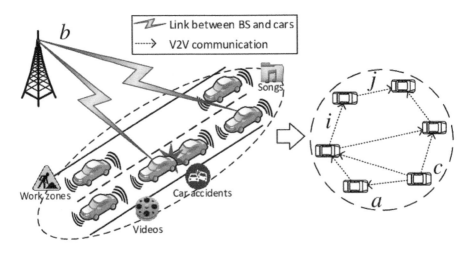

Figure 11.2. *V2V via 5G. For a color version of this figure, see www.iste.co.uk/haddadou/edge.zip*

The third solution is a direct communication between vehicles because of light beams that obviously do not cross obstacles. This technology is generally called VLC (Visible Light Communication) with products such as Li-Fi. Figure 11.3 illustrates this last solution.

Figure 11.3. *Communication by light between two vehicles. For a color version of this figure, see www.iste.co.uk/haddadou/edge.zip*

Another major difference between the three solutions is where the control data is processed. In the case of vehicle-to-vehicle communications via Wi-Fi, the control takes place in servers located in the vehicles that are in-vehicle data centers. For techniques around 5G, the control is achieved in the MEC data center. Finally, when communications are carried out via light, control is carried out directly from the server to the control server.

We will take a closer look at the three solutions we described earlier and detail the virtualization in the three cases: control, data network and diagnostic network.

11.1. ETSI ITS-G5

The ever-increasing road traffic generates numerous problems of congestion, safety and environmental impact. Information and communication technologies provide solutions to these problems by automating them and making vehicles autonomous. Intelligent Transport Systems (ITS) encompass a wide variety of solutions to achieve vehicle-to-vehicle communications. These solutions are designed to increase travel safety, minimize environmental impact, improve traffic management and, more generally, optimize travel.

Driver assistance for autonomous vehicles helps them maintain a prescribed speed and distance, drive in their lane, avoid overtaking in dangerous situations and cross intersections safely, which has positive effects on safety and traffic management. However, the benefits could be further amplified if vehicles were able to continuously communicate with each other or with the road infrastructure. We will first look at the important technical elements in the first generation of ITS.

The first component toward automation comes from the radar technology that is developed to be used in vehicles, to smooth traffic and avoid accidents. Several types of radar can be explained, including long-range radars that operate at 77 GHz and are used, for example, in speed control between following vehicles. This allows a vehicle to maintain a constant cruising distance with the vehicle in front. Short-range radars are used to manage collision avoidance systems that operate at 24 and 79 GHz. In the event that a collision is unavoidable, the vehicle has time to prepare for the collision by braking, of course, but also by activating protective equipment such as airbags to minimize injuries to passengers and others. A band has been allocated on 79 GHz to implement this radar service.

In recent years, research on intelligent vehicles has focused on cooperative ITS (C-ITS) in which vehicles communicate with each other and/or with the infrastructure. A C-ITS can significantly increase the quality and reliability of the information available in the vehicles, their location and the road environment. A C-ITS enhances existing services and creates new ones for road users, bringing significant social and economic benefits, increased transport efficiency and improved safety.

The technology used to enable vehicles to communicate with each other comes from the IEEE 802.11p standard, which is an approved amendment to the IEEE 802.11 standard to add wireless access for vehicular environments (WAVE) defining an in-vehicle communication system. It defines the enhancements to the 802.11 standard (the working group that defines Wi-Fi standards) required to support ITS applications. This includes data exchange between high-speed vehicles and between vehicles and roadside infrastructure, known as vehicle-to-infrastructure (V2I) communication, in the ITS band, licensed at 5.9 GHz (5.85-5.925 GHz). IEEE 1609 is the standard that applies to the upper layer based on IEEE 802.11p. It is also the basis for a European standard for vehicular communication known as ETSI ITS-G5.

The IEEE 802.11p standard uses the Enhanced Distributed Channel Access (EDCA) MAC channel access method, which is an enhanced version of the Distributed Coordination Function (DCF) that is the core solution for Wi-Fi defined by the 802.11 group. EDCA uses Carrier Sense Multiple Access (CSMA) with CSMA/Collision Avoidance (CSMA/CA), which means that a node that wants to transmit must first detect whether the medium is free or not. If it is free during the AIFS (Arbitration Inter-Frame Space) time, it defers the transmission by drawing a random time. The procedure for determining the random time works is as follows:

– The node draws a random value uniformly over the interval [0, CW + 1] where the initial value CW (containment window) is equal to CWmin.

– The size of the backoff interval doubles each time the transmission attempt fails, until the CW value is equal to CWmax. If the backoff value has the value 0, the sender immediately sends its packet. To ensure that urgent messages are transferred with high security, the 802.11p protocol uses priorities corresponding to different access classes (AC). There are four classes of data traffic with different priorities: Background traffic (BK or AC0), Best Effort traffic (BE or AC1), Video traffic (VI or AC2) and Voice traffic (VO or AC3). Different AIFS – wait time before transmitting – and CW are chosen for the different access classes. This algorithm is quite similar to those of other Wi-Fi networks and allows, at a very low cost, to obtain a good quality of service for priority clients and a lower quality of service for non-priority data.

To conclude this first section, G5 is the most advanced of all the proposed techniques for interconnecting vehicles and building a vehicular network. Numerous experiments to evaluate its performance have been conducted in the field. This solution is perfectly capable of supporting inter-vehicle communications and represents a simple method to implement with a relatively low cost. What can be reproached to this solution is that it is limited to the control part because of the

Wi-Fi channel used which has a very limited bandwidth. The CSMA/CA solution can also be criticized for being somewhat random, and if the number of vehicles is large, strong interference can significantly alter performance.

11.2. 5G standardization

The standardization of 5G is almost complete. The first phase represented by release 15 was completed in 2018, the second phase in 2020 and the third phase, release 17, in June 2022. The first two phases are important for vehicle control. Indeed, the first phase specified the radio part and the second phase the critical missions. The latency retained in this second phase of the standardization is 1 ms, which is a perfectly acceptable time for vehicle control. The reaction time of an individual's foot to brake being greater than 100 ms, it is considered that the reaction times taking into account the latency and control calculations made in a server should be less than 10 ms. The third phase released in June 2022 brings complements and especially a complete definition of 5G services.

11.2.1. *5G vehicular networks*

For over 15 years, the ITS (Intelligent Transportation System) has defined a spectrum band, harmonized worldwide in the 5.9 GHz band, dedicated to vehicle traffic safety applications. The IEEE 802 11p standard has been chosen as we have just seen for the communication system providing active safety functions. Over the last 15 years, authorities, industry and road operators have managed to deploy this common communication infrastructure using this standard, but on a relatively small scale. An alternative comes from a rather similar technique, this time based on the standardization carried out by 3GPP for mobile networks, offering direct communication between vehicles. This alternative to traditional IEEE 802 11p communication can combine the strengths of cellular connectivity based on V2N communication with the strengths of direct V2V communication. In the automotive industry, since cellular connectivity is required for various other controls, standardized direct communication can be established in parallel with the deployment of cellular connectivity.

V2V (Vehicle-to-Vehicle) or V2I (Vehicle-to-Infrastructure) are appropriate solutions to manage safety elements in the vehicle domain. Many manufacturers have already implemented many use cases of these technologies by adding smart sensors in the vehicle and collecting information during V2N (Vehicle-to-Network) communications. We will now describe some examples in this area.

A first example is a collision warning based on sensor information from vehicles, pedestrians or the infrastructure indicating the position and speed of vehicles that may be involved in the collision. A second example is the detection of a dangerous location from a back-office service, such as the detection of a strong deceleration of a vehicle during emergency braking, reported by sensors from different vehicles. While a safety function relies on data from other vehicles and needs to provide active safety, the network latency must be strongly limited to a maximum time of 100 ms. Without this upper bound, information must be transmitted directly between vehicles, to or from the road infrastructure, or to or from pedestrians. A third example is given by V2I: for example, a traffic light transmits to vehicles when the light is about to turn green. Another example is a construction vehicle transmitting information about speed limits around the construction site or about the dangers of the work in progress.

The new term used in the 5G world to describe communications in the vehicular environment is C-V2X (Cellular-V2X). The basic V2X mode includes two communication modes: mode 3 and mode 4. In mode 3, the cellular network allocates and manages the radio resources used by the vehicles for their communications. Signaling is routed through the eNodeB or through the MEC datacenter that virtualizes the eNodeB. In mode 4, the vehicles autonomously select the radio resources to perform direct V2V communications, directly from one vehicle to another. Mode 4 uses scheduling for congestion control, which allows priority packets to transit first and give them the required QoS. V2X communications in mode 4 can operate without cellular coverage, going directly from one vehicle to another if they are close enough to each other to be within their respective coverage areas. Mode 4 is considered the basic mode of V2V, as traffic safety applications cannot depend on the availability of radio coverage. These control applications must be able to run without coverage outside of the antennas in the vehicles.

In the event that a cellular connection cannot be established, direct communication via the PC5 interface (the name of the interface from one vehicle to another) must work to maintain communication. This requirement comes from the fact that vehicles may not be covered by a radio cell associated with an eNodeB or vehicles are using their V2N connectivity from the Uu interface for other services. The ecosystem based on the PC5 interface must therefore be able to function at all times, regardless of the mobile operator network to which the connected vehicle belongs. In addition to all the technical reasons, direct communication is a matter of trust because, unlike a telecom operator, no third party is involved in establishing the communication. If a vehicle cannot trust the information transmitted via PC5 communication, the system cannot work. The security systems for direct communication are largely supported by the C-V2X standard. All security and data privacy rules are defined in the C-V2X direct communication standard, just as they are specified in the IEEE 802.11p standard. On the other hand, these security rules benefit from V2N functionalities for operational purposes, such as distribution, calculation, storage, certificate renewal and revocation lists. For this reason, C-V2X emphasizes direct PC5 communication with different service classes:

– different types of user equipment involved in C-V2X (vehicles, pedestrian areas, roadside infrastructure);

– different network providers, including lack of network coverage;

– different components of the safety management system that establish a level of trust between the entities of the system.

11.2.2. *C-V2X technology overview*

In 2015, 3GPP specified V2X features to support LTE-based V2X services in release 14. V2X covers V2V, V2I, vehicle-to-pedestrian (V2P) and V2N specifications. The first infrastructure connection specifications occurred in September 2016. Additional enhancements to support new V2X operational scenarios were added in release 14, which was finalized in June 2017. The various types of V2X communications continued to be enhanced in all three phases of 5G, culminating in a full specification with 3GPP release 17, released in June 2022.

V2V communications are based on D2D device-to-device communications defined in the Proximity Services (ProSe) framework, which are defined in releases 12 and 13. As part of these services, a new D2D interface, named PC5, which we have already discussed in the previous section, was introduced to support the use case of inter-vehicle communications, in particular to define the enhancements needed to support V2X requirements for inter-vehicle communication. Enhancements were made to handle high vehicle speeds (Doppler shift/frequency shift, etc.) up to 500 km/h, synchronization outside of eNodeB coverage, improved resource allocation, congestion control for high traffic load and traffic management for V2X services. Two modes of operation for LTE-V V2V communications have been introduced in release 14: the first corresponds to communication via the 4G infrastructure and the second corresponds to communication over the PC5 (V2V) peer-to-peer interface.

The Uu interface corresponding to V2N represents the case where communication is implemented via an eNodeB. 5G V2V communication over the PC5 interface is supported via two modes: managed mode (PC5 mode 3), which operates when communication between vehicles is scheduled by the network, and unmanaged mode (PC5 mode 4), which operates when vehicles communicate independently of the network. PC5 mode 4 and interference management are supported by distributed algorithms between vehicles, while PC5 mode 3 scheduling and interference management are managed via the base station (eNodeB) by control signaling using the Uu interface. Deployment scenarios, supported by one or more operators, benefit from the additional advantages of using Uu-based connectivity with broadcast services.

C-V2X involves V2N/V2N2V/V2I/V2N2P connectivity based on a so-called Uu interface with V2V/V2I/V2P connectivity based on PC5 (reference point for a direct vehicle-to-vehicle connection). We will mainly discuss the advantages of PC5 technology for V2X, also regarding how to deploy it. Furthermore, this technology takes into account all the aspects of V2N, which must be present in the PC5 direct connection in order to obtain a perfectly controlled connectivity. Operationally, the PC5 interface can be established for the following communications:

– vehicle to vehicle (V2V);

– vehicle to infrastructure (V2I);

– between a vehicle and road users, whether pedestrians, cyclists or other road users (V2P).

From a physical layer perspective, signals are transmitted using LTE technology. LTE defines an elementary resource by a frequency band of 15 kHz taking a time of 0.5 ms. These elementary resources are grouped into elementary blocks of data to be transmitted consisting of 12 bands of 15 kHz following the technique used in OFDM. An elementary block in LTE uses a spectrum of 12×15 kHz = 180 kHz. It is called RB (Resource Block), which covers all 12 bands of 15 kHz.

In LTE-V (LTE-Vehicle), the frame lasts 10 ms. It is divided into 10 sub-frames of 1 ms duration. Each subframe is divided into two slots of 0.5 ms. A slot therefore lasts 0.5 ms, during which seven symbols are transmitted per OFDM band. Since there are 12 sub-bands, this gives a total of 84 symbols that are transmitted every 0.5 ms. In V2X, we do not consider splitting the slot into two parts. For a 1 ms slot, 168 symbols are carried. A symbol can transmit from 1 bit to 6 bits, depending on the chosen modulation (QPSK at 16 QAM in V2X, but this can go up to 128 QAM in classical LTE).

5G NR V2X works in a similar way with some differences to improve, among other things, the latency time and reach values perfectly adapted for control in the vehicular world.

Transmissions in NR V2X use OFDM as in 4G but with a cyclic prefix (CP). The frame structure is organized in radio frames (also called simply, frames), each with a duration of 10 ms. A radio frame is divided into 10 sub-frames, each with a duration of 1 ms. The number of slots per sub-frame and the subcarrier spacing (SCS) for the OFDM waveform can be flexible for NR V2X. To support various requirements and different operating frequencies in FR1 (band between 410 and 7125 Mhz) and FR2 (band between 24.250 and 52.600 GHz), an OFDM numerology has been chosen for NR V2X based on release 15. Each OFDM numerology is defined by an SCS and a CP. NR V2X supports multiples of 15 kHz (i.e. the SCS in LTE V2X) for the SCS of the OFDM waveform given by $2^n \times$ 15 kHz, where n is an SCS configuration factor. For NR V2X, the SCS configuration factor can be $n = 0, 1, 2, 3$ such that the SCS can be equal to 15, 30, 60 or 120 kHz. In FR1, 15, 30 and 60 kHz are supported for SCS, while 60 and 120 kHz are supported for SCS in FR2. The higher SCS support improves the robustness of the OFDM waveform. On the other hand, there is a degradation from an accentuated Doppler effect and an offset in the carrier frequency.

11.3. Visible light communication

Light is another means of communication that is very useful in vehicular networks: between two cars following each other, there is no obstacle that would stop the light. On the other hand, the communication can only be achieved from vehicle to vehicle, and the vehicular network is realized by point-to-point links that must be relayed by equipment located in the vehicles. In this section, we will look at the characteristics of these networks, which are called VLCs, and one of the major products is Li-Fi.

Li-Fi is a wireless communication technology based on the use of visible light, with a wavelength between 460 and 670 THz. The principle of Li-Fi is based on the coding and sending of data by an amplitude modulation of light, according to a well standardized protocol. The protocol layers of Li-Fi are adapted to wireless communications up to 10 m, which is an acceptable limit for communication between two vehicles.

The data are coded to allow a crossing of the optical channel which is far from perfect. The signal in electrical form is converted into a light signal by means of an electronic circuit that allows the light intensity of the LEDs to vary according to the data to be transmitted. The modulation used is an intensity modulation where the logical 0's and 1's are transmitted according to the Manchester coding. The emitted light propagates in the environment and undergoes deformations for various reasons such as weather conditions. This environment and the associated distortions are grouped under the term optical channel.

Li-Fi is a wireless technology that uses a protocol very similar to those of the 802.11 group, but it uses communications using light instead of radio frequency waves. Visible and invisible light uses a much larger bandwidth than radio frequencies, which obviously results in a much higher throughput. Specifically, Li-Fi has followed the IEEE 802.15.7 standard, which has been modified to reflect the latest advances in optical wireless communications. In particular, the standard now incorporates modulation methods with optical orthogonal frequency division multiplexing (O-OFDM). IEEE 802.15.7 also defines the physical layer (PHY) and the medium access control (MAC) layer. The standard defines data rates that allow audio, video and multimedia services to be carried without difficulty. It takes into account the mobility of the optical transmission, its compatibility with the artificial lighting present in the infrastructures and the interferences that can be generated by the ambient lighting. The standard defines three PHY layers with different data rates:

– PHY 1 is designed for outdoor applications and operates from 11.67 to 267.6 kbps;

– PHY 2 layer enables reaching data rates ranging from 1.25 to 96 Mbps;

– PHY 3 is used for many transmission sources with a particular modulation method called CSK (Color Shift Keying). PHY 3 can offer data rates ranging from 12 to 96 Mbps.

The recognized modulation formats for PHY 1 and PHY 2 use OOK (On-Off Keying) and VPPM (Viable Pulse Position Modulation). The Manchester coding used for PHY 1 and PHY 2 layers includes the clock in the transmitted data by representing a logical 0 with an OOK symbol "01" and a logical 1 with an OOK symbol "10", all with a DC component. The continuous component avoids the extinction of the light in the case of an extended series of logical 0.

11.4. The architecture of vehicular networks

The architecture of a vehicular network is mainly composed of antennas, since mobility must be supported. The communications take place between two antennas, resulting in V2V, V2P, V2N and vehicle-to-roadside communications (V2RC) connections. The antennas are mainly 5G antennas, since the automatic driving processes are positioned in the operators' data centers integrating MEC data centers.

The virtual machines in these data centers can of course belong to roadside operators. Figure 11.4 provides an example of a vehicular network architecture.

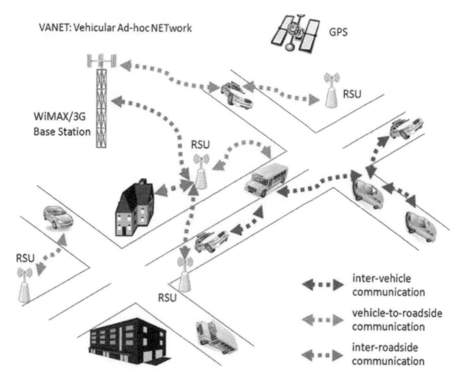

Figure 11.4. *Connections in a vehicular network. For a color version of this figure, see www.iste.co.uk/haddadou/edge.zip*

Cloud Continuum plays an important role in positioning virtual machines in the best locations as needed. In particular, when a particularly short latency is required, data centers located in the vehicle should be used, for example, for emergency braking. Many control processes are located in MEC data centers, which become indispensable due to the latency time of no more than 1 ms. This value is necessary for many control applications. Finally, the cloud is used for non-real-time services that require significant storage and memory resources.

Figure 11.5 illustrates the uses of the Edge and the Cloud.

192   Cloud and Edge Networking

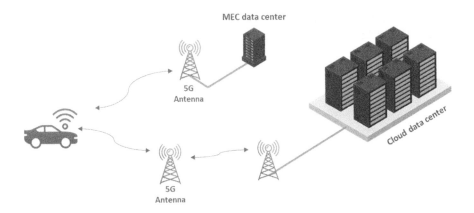

Figure 11.5. *The Cloud for vehicular networks. For a color version of this figure, see www.iste.co.uk/haddadou/edge.zip*

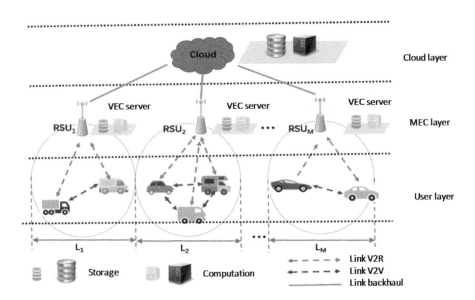

Figure 11.6. *The Edge and the Cloud in a vehicular environment. For a color version of this figure, see www.iste.co.uk/haddadou/edge.zip*

Vehicular world functions are also dispersed across the Cloud Continuum, and a representation of application positioning is available in Figure 11.6. Real-time functions as well as application services are shown at the top of the figure, while infrastructure and communications take up space toward the bottom of the figure.

Vehicular networks will be a particularly important category of networks in the coming years as driving becomes automated. As a result, drivers will be free to connect to the Internet for services relevant to work, home or any other desired environment.

11.5. Conclusion

Vehicular networks are expanding rapidly. Automated driving at somewhat higher speeds requires coordination that can only be achieved by connecting vehicles to each other. The market being extremely important, the industry is preparing for a battle to impose their standard. The first parties come from the G5 ETSI ITS-G5 standard using Wi-Fi IEEE 802.11p, and many tests have already been carried out showing the effectiveness of this method. However, this solution can see its speeds drop in poor conditions, especially with interference between different antennas. 5G, which is a much younger technology, is arriving in force with synchronous techniques from the world of telecom operators. This solution is seen as a potential winner because of its synchronous nature and the quality of service it brings, especially for the security part. Moreover, the 5G world has specified a particular mode, which allows the implementation of a direct communication from vehicle to vehicle without an eNodeB box. This direct communication solution will be used when vehicles are no longer under the coverage of a large 5G antenna.

11.6. References

Abbas, N., Zhang, Y., Taherkordi, A., Skeie, T. (2018). Mobile edge computing: A survey. *IEEE Internet of Things Journal*, 5(1), 450–465.

Abbas, F., Gang, L., Pingzh, F., Zahid, K. (2020). An efficient cluster based resource management scheme and its performance analysis for V2X networks. *IEEE Access*, 8, 87071–87082.

Allan, C., Pacheco, L., Rosário, D., Villas, L., Loureiro, A., Sargento, S., Cerqueira, E. (2020). Skipping-based handover algorithm for video distribution over ultra-dense vanet. *Comput. Netw.*, 176, 107252.

Arena, F. and Pau, G. (2019). An overview of vehicular communications. *Future Internet*, 11(2), 27–39.

Baron, B., Campista, M., Spathis, P., Costa, L.H.M.K., Amorim, M.D., Duarte, O.C.M.B., Pujolle, G., Viniotis, Y. (2016). Virtualizing vehicular node resources: Feasibility study of virtual machine migration. *Vehicular Communications*, 4(C), 39–46.

Cao, J., Maode, M., Hui, L., Ruhui, M., Yunqing, S., Pu, Y., Lihui, X. (2029). A survey on security aspects for 3GPP 5G networks. *IEEE Commun. Surv. Tutor.*, 22, 170–195.

Chen, M., Li, W., Hao, Y., Qian, Y., Humar, I. (2018). Edge cognitive computing based smart health care system. *Future Generation Computer Systems*, 86, 403–411.

Cui, J., Wei, L., Zhang, J., Xu, Y., Zhong, H. (2018). An efficient message-authentication scheme based on edge computing for vehicular ad hoc networks. *IEEE Transactions on Intelligent Transportation Systems*, 20, 1621–1632.

Faris, H. and Setiadi, Y. (2019). Development of communication technology on VANET with a combination of ad-hoc, cellular and GPS signals as a solution traffic problems. In *7th International Conference on Information and Communication Technology*, IEEE, Kuala Lumpur, Malaysia.

Gupta, S. and Chakareski, J. (2020). Lifetime maximization in mobile edge computing networks. *IEEE Transactions on Vehicular Technology*, 69(3), 3310–3321.

Hakeem, S.A., Anar, H., Hyung, K. (2020). 5G-V2X: Standardization, architecture, use cases, network-slicing, and edge-computing. *Wireless Networks*, 26, 6015–6041.

Hatoum, A., Langar, R., Aitsaadi, N., Boutaba, R., Pujolle, G. (2014). Cluster-based Resource Management in OFDMA Femtocell Networks with QoS Guarantees. *IEEE Transactions on Vehicular Technology*, 63(5), 2378–2391.

Hwang, H. and Young-Tak, K. (2020). Management of smart vehicular handovers in overlapped V2X networks NOMS 2020. In *IEEE/IFIP Network Operations and Management Symposium*, IEEE, Budapest, Hungary.

Iacovos, I., Vassiliou, V., Christophorou, C., Pitsillides, A. (2020). Distributed artificial intelligence solution for D2D communication in 5G network. *IEEE Syst. J.*, 14, 4232–4241.

Ikram, A., Lawrence, T., Li, F. (2020). An efficient identity-based signature scheme without bilinear pairing for vehicle-to-vehicle communication in VANETs. *J. Syst. Archit.*, 103, 101692.

Jawad Kadhim, A. and Naser Jaber, I. (2021). Toward electrical vehicular ad hoc networks: E-VANET. *J. Electr. Eng. Technol.*, 16, 1667–1683.

Khedr, W., Hosny, K.M., Khashaba, M.M., Amer, F.A. (2020). Prediction-based secured handover authentication for mobile cloud computing. *Wirel. Netw.*, 26, 4657–4675.

Khoder, R., Naja, R., Mouawad, N., Tojme, S. (2020). Vertical handover network selection architecture for VLC vehicular platoon driving assistance. In *IEEE 31st Annual International Symposium on Personal, Indoor and Mobile Radio Communications*, IEEE.

Kim, Y., An, N., Park, J., Lim, H. (2018). Mobility support for vehicular cloud radio-access-networks with edge computing. In *7th International Conference on Cloud Networking (CloudNet)*, IEEE.

Kozik, R., Choras, M., Ficco, M., Palmieri, F. (2018). A scalable distributed machine learning approach for attack detection in edge computing environments. *Journal of Parallel and Distributed Computing*, 119, 18–26.

Kumar Pranav, S., Sunit Kumar, N., Sukumar, N. (2019). A tutorial survey on vehicular communication state of the art, and future research directions. *Vehicular Communications*, 18, 100164.

Langar, R., Secci, S., Boutaba, R., Pujolle, G. (2015). An operations research game approach for resource and power allocation in cooperative femtocell networks. *IEEE Transactions on Mobile Computing*, 14(4), 675–687.

Li, X., Dang, Y., Aazam, M., Peng, X., Chen, T., Chen, C. (2020). Energy efficient computation offloading in vehicular edge cloud computing. *IEEE Access*, 8(37), 632–644.

Livinus, T., Ayaida, M., Tohme, S., Afilal, L.-E. (2020). A mobile internal vertical handover mechanism for distributed mobility management in VANETs. *Vehicular Communications*, 26, 100177.

Martín-Sacristán, D., Sandra, R., Roger David, G., Monserrat, J.F., Panagiotis, S., Chan, Z., Kaloxylos, A. (2020). Low-latency infrastructure-based cellular V2V communications for multi-operator environments with regional split. *IEEE Trans. Intell. Transp. Syst.*, 22, 1052–1067.

Mendiboure, L., Chalouf, M., Krief, F. (2020). Survey on blockchain-based applications in internet of vehicles. *Comput. Electr. Eng.*, 84, 106646.

Monteiro, T., Pellenz, M.E., Penna, M.C., Enembreck, F., Souza, R.D., Pujolle, G. (2012). Channel allocation algorithms for WLANs using distributed optimization. *International Journal of Electronics and Communications*, 66(6), 480–490.

Movahedi, Z., Ayari, M., Langar, R., Pujolle, G. (2012). A survey of autonomic network architectures and evaluation criteria. *IEEE Communications Surveys and Tutorials*, 14(2), 491–513.

Ndashimye, E., Nurul, S.I., Sayan Kumar, R. (2021). A multi-criteria based handover algorithm for vehicle-to-infrastructure communications. *Comput. Netw.*, 185, 107652.

Nyangaresi Omollo, V., Anthony Joachim, R., Silvance Onyango, A. (2020). Efficient group authentication protocol for secure 5G enabled vehicular communications. *16th International Computer Engineering Conference, ICENCO*, IEEE Proceedings IEEE, New York.

Pachat, J., Nujoom Sageer, K., Deepthi, P.P., Rajan, S. (2020). Index coding in vehicle to vehicle communication. *IEEE Trans. Veh. Technol.*, 69, 11926–11936.

Pacheco, L., Iago, M., Hugo, S., Helder, O., Denis, R., Cerqueira, E., Neto, A. (2019). A handover algorithm for video sharing over vehicular networks. In *9th Latin-American Symposium on Dependable Computing*, IEEE.

Pang, P., Sun, L., Wang, Z., Tian, E., Yang, Z. (2015). A survey of cloudlet based mobile computing. In *Proceedings of the International Conference on Cloud Computing and Big Data*.

Rocha, F., Miguel, L., Zúquete, A., Sargento, S. (2020). Complementing vehicular connectivity coverage through cellular networks. In *27th International Conference on Telecommunications, ICT*, IEEE.

Rodriues, J. (2018). *Advances in Delay-tolerant Networks (DTNs): Architecture and Enhanced Performance*. Woodhead Publishing, Sawston.

Sadip, M., Roy, A., Majumder, K., Phadikar, S. (2020). QoS aware distributed dynamic channel allocation for V2V communication in TVWS spectrum. *Comput. Netw.*, 171, 107126.

Sahirul, A., Sulistyo, S., Mustika, I., Adrian, R. (2020a). Utility-based horizontal handover decision method for vehicle-to-vehicle communication in VANET. *Int. J. Intell. Eng. Syst.*, 13(13), 1–10.

Sahirul, A., Sulistyo, S., Wayan, M.I., Adrian, R. (2020b). Fuzzy adaptive hysteresis of RSS for handover decision in V2V VANET. *Int. J. Commun. Netw. Inf. Secur.*, 12, 433–439.

Sailendra, K. and Barooah, M. (2020). Lifi based scheme for handover in VANET: A proposed approach. *International Conference on Information, Communication and Computing Technology*, Springer, Singapore.

Sethom, K. and Pujolle, G. (2018). Spectrum mobility management in cognitive two-tier networks. *International Journal of Network Management*, 28(3), e2019.

Sharma, A. (2020). Mission swachhta: Mobile application based on mobile cloud computing. In *2020 10th International Conference on Cloud Computing, Data Science Engineering*.

Singh, K., Sahil, S., Sunit Kumar, N., Sukumar, N. (2019). Multipath TCP for V2I communication in SDN controlled small cell deployment of smart city. *Vehicular Communications*, 15, 1–15.

Sliwa, B., Falkenberg, R., Wietfeld, C. (2020). Towards cooperative data rate prediction for future mobile and vehicular 6G networks. In *Proc. 6G Wireless Summit*.

Storck, C., Efrem, E., Guilherme, O.G., Raquel, M., Duarte-Figueiredo, F. (2021). FiVH: A solution of inter-V-Cell handover decision for connected vehicles in ultra-dense 5G networks. *Vehicular Communications*, 28, 100307.

Tang, F., Kawamoto, Y., Kato, N., Liu, J. (2020). Future intelligent and secure vehicular network toward 6g: Machine-learning approaches. *Proc. IEEE*, 108, 292–307.

Tokarz, K. (2019). A review on the vehicle to vehicle and vehicle to infrastructure communication. *International Conference on Man–Machine Interactions*, Springer, Cham.

Tong, W., Azhar, H., Wang, B., Sabita, M. (2019). Artificial intelligence for vehicle-to-everything: A survey. *IEEE Access*, 7, 10823–10843.

Wang, S., Zhao, Y., Huang, L., Xu, J., Hsu, C.-H. (2017). QoS prediction for service recommendations in mobile edge computing. *Journal of Parallel and Distributed Computing*, 127, 134–144.

Wu, Y. and Hu, F. (2017). *Big Data and Computational Intelligence in Networking*. CRC Press, Boca Raton.

Xiying, F., Di, L., Bin, F., Shaojie, W. (2021). Optimal relay selection for UAV-assisted V2V communications. *Wirel. Netw.*, 27, 3233–3249.

Yao, D., Yu, C., Yang, L.T., Jin, H. (2019). Using crowdsourcing to provide QoS for mobile cloud computing. *IEEE Transactions on Cloud Computing*, 7(2), 344–356.

Zhang, Z., Xiao, Y., Ma, Z., Xiao, M., Ding, Z., Lei, X., Karagiannidis, G.K., Fan, P. (2019a). 6G wireless networks: Vision, requirements, architecture, and key technologies. *IEEE Vehicular Technology Magazine*, 14(3), 28–41.

Zhang, Z., Zhang, W., Tseng, F. (2019b). Satellite mobile edge computing: Improving QoS of high-speed satellite-terrestrial networks using edge computing techniques. *IEEE Network*, 33(1), 70–76.

Zhang, J., Guo, H., Liu, J., Zhang, Y. (2020). Task offloading in vehicular edge computing networks: A load-balancing solution. *IEEE Transactions on Vehicular Technology*, 69(2), 2092–2104.

Zhao, Z., Min, G., Gao, W., Wu, Y., Duan, H., Ni, Q. (2018). Deploying edge computing nodes for large-scale IoT: A diversity aware approach. *IEEE Internet of Things Journal*, 5(5), 3606–3614.

Zhifeng, Y., Yihua, M., Yuzhou, H., Weimin, L. (2020). High-efficiency full-duplex V2V communication. *2nd 6G Wireless Summit. 6G SUMMIT*, IEEE.

12

The Cloud Continuum and Industry 4.0

Transformations in the industrial world happen in spurts when a new revolutionary solution comes along. We are at the fourth revolution with Industry 4.0. Figure 12.1 lists these different stages to arrive at the fourth one, whose objective is to totally automate industrial production from the product order to its delivery.

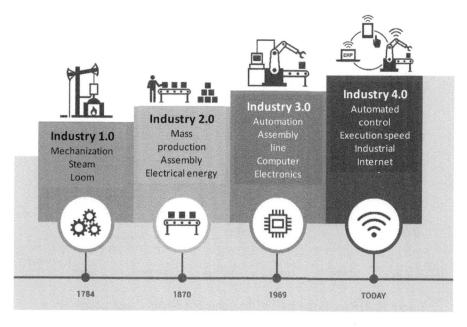

Figure 12.1. *Revolutions in the industrial world. For a color version of this figure, see www.iste.co.uk/haddadou/edge.zip*

The revolution brought about by Industry 4.0 comes from the mutations due to digital technology. The control of industrial equipment, such as robots or smart sensors, falls into this category, where real-time and remote control are paramount. This industrial revolution is based on automated systems that can be activated in real time. Many important changes are needed in manufacturing, engineering, use of new materials, supply chain and lifecycle management. Smart factories must be particularly flexible and automated. Services and applications must be combined with platforms that enable the automation of the environment, which will also be used for communication between people, objects and systems. Latency requirements for different applications in the automated manufacturing world range from tens of milliseconds for mechanical, to milliseconds for M2M (Machine-to-Machine) communications, to 1 ms for electrical devices. Reaction times for control cycles of fast-moving devices can reach values well below 1 ms. Finally, some highly sensitive subsystems require even shorter latencies, in the order of tenths of a millisecond.

Figure 12.2 describes the latency and throughput required for a range of applications with the new capabilities brought by 5G. These include massive real-time gaming, multi-user video conferencing, real-time virtual reality, remote office work, etc. Industry 4.0 is in the new area brought by 5G with a response time of the order of a millisecond but with various speeds ranging from small to particularly high.

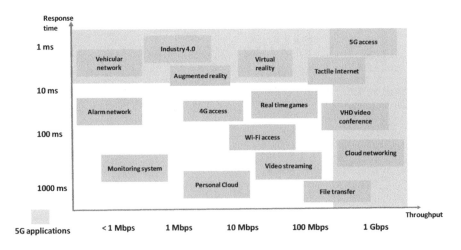

Figure 12.2. *Response time and throughput of different applications. For a color version of this figure, see www.iste.co.uk/haddadou/edge.zip*

12.1. The features needed to achieve Industry 4.0

The functionalities required to achieve an Industry 4.0 environment are numerous. Several types of networks are competing to enter this huge market. The two main networks are the wired industrial networks (Ethernet technologies) and the wireless technologies – especially using 5G to bring about synchronization. The rise of the 5G solution is being closely watched by the industrial sector because of its performance on millisecond latency and antenna directivity which allows synchronization of machine tools and robots by transmitting synchronous streams in parallel using directional antennas.

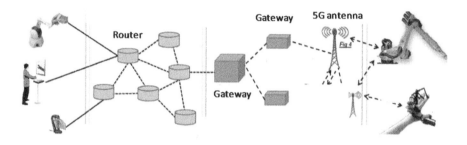

Figure 12.3. *Architecture of an Industry 4.0 application. For a color version of this figure, see www.iste.co.uk/haddadou/edge.zip*

A 5G-driven communication architecture, consisting of the radio interface to connect sensors, actuators, and more generally Industrial Internet of Things (IIoT) objects and the Radio Access Network (RAN), meets the requirements for realizing an Industry 4.0 environment. As shown in Figure 12.3, the end-to-end architecture for Industry 4.0 is divided into three distinct domains: the master domain, the network domain and the controlled domain. The master domain contains a human or machine and a system interface available to the human or automated operator. This interface contains a control system that converts the operator's input into an Industry 4.0 input, using various coding techniques. If the control device is a haptic device, it enables a human to touch, feel, manipulate or control objects in real or virtual environments. The control device primarily drives the operation of the controlled domain. In the case of a networked control system, the master domain that contains the operator has a controller that drives the sensors and actuators in the network. The controlled domain consists of an object that can be a robot, a machine tool or any other machine. This object is controlled directly by the master domain via various control signals to interact on the remote environment. In case of remote operation with haptic feedback, energy is exchanged between the operator and the controlled domains, forming a global control loop.

Networks provide the basic support that is absolutely necessary for communication between the machine or human and the industrial equipment controlled remotely. This network must have all the necessary resources to allow for quality communication with very short latency times meeting the real time required for remote machine control. Carrier grade networks and in particular 5G networks provide these properties. The carrier grade indicates networks with a very high availability, quality of service and a system to verify this quality. In order to contain all the necessary functionalities to realize a tactile network, whether it is 5G or more generally a carrier grade network, it is necessary that the following properties are verified:

– Highly responsive network connectivity: this property is particularly important for systems with critical constraints, for example, synchronized robot-to-robot communication for real-time control and automation of dynamic processes in industrial automation or manufacturing of complex objects.

– Carrier grade access reliability and availability: network connectivity must be ultra-reliable, which is an Industry 4.0 requirement. However, the reliability requirement is very specific to the application process and can vary greatly depending on the desired outcome.

– The possibility to manage multi-access: the simultaneous use of several networks, or several directional antennas, is important to increase the reliability and availability of the network. This also represents a solution to reduce latency with controllers that are able to choose the optimal network or antenna at any time, both for its latency and for the security of the communication. The controller must be equipped with sufficient intelligence to make choices in real time, directing the flows to the network that has the required performance for the application to run smoothly. Artificial intelligence (AI) techniques should play a crucial role in automated industrial manufacturing.

– Proactive allocation of resources to achieve the required quality of service: resource control has a direct impact on throughput, latency, reliability, quality of service and protocol layer performance. To achieve very strict latency, resources must be prioritized for Industry 4.0 critical applications. In addition, when applications need to coexist, networks require strong intelligence to achieve flexible resource control and be able to provide certain functionality on demand. In particular, for 5G which is one of the main networks used for Industry 4.0, radio frequency allocation is a crucial task to avoid interference between antennas. The access network – the RAN – must also provide the necessary quality of service.

RAN virtualization technologies can go a long way but can also increase response times if the architecture in place does not meet real-time needs.

– Feedback loop control: feedback loops react on systems that deviate from the customized operation of an application and are necessary to maintain the desired quality of service. Industry 4.0 processes require such feedback loops when signals are exchanged bidirectionally over the network, to enable remote control of real or virtual objects and systems.

– Global design of the communication infrastructure: indeed, Industry 4.0 supports end-to-end applications, that is, from the control software to the machine tool, and therefore requires that the sum of the systems crossed is always able to offer the necessary service qualities. Each component of the overall system must have the necessary resources to achieve the quality of service required for Industry 4.0, but their juxtaposition may no longer achieve the performance required for goods production applications to operate in real time, possess the right security or achieve the availability of a carrier grade environment. Therefore, it is important to design and size all the technology components involved in the control network and to consider response time, throughput, security and availability of the whole. If the end-to-end network is 5G, the overall review requires proper sizing of the radio access and RAN, which can be easier, especially if the whole thing is orchestrated, controlled and managed by the same system.

12.2. Technical specifications for 5G

To meet the challenges imposed by the demands of Industry 4.0 applications, technology innovations are needed in many aspects of network design, from the physical layer and protocols to resource management. When it comes to 5G, there is a need to be able to manage radio resources and the RAN. Essentially, the technology faces the challenge of providing sufficient network connectivity to enable the latency and reliability required for real-time physical interaction. While some Industry 4.0 applications such as cloud-associated robotics or machine tool control simultaneously require low latency and high reliability, other Industry 4.0 applications rely primarily on low-latency communication where security may be of a lower order.

We can think of many measures that can help reduce latency and increase the reliability of mobile networks. First, we need to reduce latency to a few milliseconds. Various means are being implemented to reduce latency in data transmissions, specifically in the next generation of 5G networks. One approach that is gaining traction is to define flexible frame structures and number these frames so that subcarrier spacings and transmission times can have different values. This allows applications with distinct requirements to be supported. Recent developments by 3GPP are moving in this direction, agreeing on much smaller base times and subcarrier spacing, for example, 0.125 ms, or even less. With the concept of access and mini-intervals (two or more OFDM symbols), it is even possible to significantly reduce latency to sub-ms values.

Limited by the speed of light, the distance between the transmitter and receiver must be as short as possible to ensure very low latency communications. Thus, if the calculations are performed in a Cloud infrastructure such as Fog or MEC, directly at the edge of the mobile network, the latency can be very small and acceptable for real-time applications. The technologies used in this virtualized environment are SDN (Software-Defined Networking) and NFV (Network Functions Virtualization). Another way to reduce latency is to support direct device-to-device communication – D2D – instead of using intermediary devices such as base stations or access points to relay data transmissions between two mobile devices.

A usual technique today to combat interference problems is to use diversity. By sending or receiving several versions of the same signal, the quality of the signal can be greatly improved, thus reducing error rates. Different types of diversity can be used: time, space and frequency diversity. Time diversity, as already mentioned, is used in retransmissions to resend data packets in case of transmission errors. In spatial diversity, micro diversity, in the form of multiple transmitting or receiving antennas, helps combat small-scale fading. In addition, macro diversity is created by using multiple base stations. Existing macro diversity solutions such as Single Frequency Network (SFN) or Coordinated Multipoint Protocol (CoMP) systems are primarily used to increase throughput and coverage at the cell edge. However, for Industry 4.0 applications requiring high reliability, these systems also help reduce transmission failures. Such macro diversity configurations involve multiple base stations that realize multi-access architectures, such that a terminal is connected to multiple antennas simultaneously. Moreover, these antennas can be heterogeneous. A third case is frequency diversity. Transmissions on different carrier frequencies are combined like Carrier Aggregation (CA) or Dual Connectivity (DC). To increase reliability, packets can be duplicated using separate radio interfaces instead of splitting the data as in CA or DC. In addition, user mobility is another source of potential poor quality due to dynamic changes in the situation. A handover occurs when a user moves from one cell to another. Even if the handover is performed correctly, an interruption time of about 50 ms occurs. If the handover fails, recovery takes much longer.

In order to avoid such interruptions, multi-access at the control plane is increasingly used. With multi-access, there are multiple paths for message delivery, even if a link is down due to an ongoing handover. This multi-access solution can also bring security to the transmission: packets of the same message go through different paths, and an attacker would have to be able to listen to the different paths and reorder the packets, which is hardly possible.

Another technique used to further improve reliability comes from network coding in which information is transmitted as multiple coded packets via different paths to the destination, thus improving reliability and latency. In summary, diversity is the main factor for improving reliability. However, one must be able to choose the best solution and type of diversity to use to meet the needs of the tactile Internet applications.

12.3. Cloud and Edge for Industry 4.0

Industry 4.0 brings together all the processes needed to automate industrial production. As shown in Figure 12.4, many elements are implemented to achieve this automation. The Internet of Things is represented by the industrial sensors and actuators that are grouped under the term IIoT (Industrial Internet of Things). These sensors can be found on machine tools or robots or cobots which are sets of robots with a strong synchronization mechanism. The control tools are located in a data center which is Fog type when it is in the company or MEC type when it is an operator who uses his/her 5G antennas to connect the sensors. The order can trace back to a Cloud when the order is not real time. Some of the ordering processes such as those coming from AI can run in the Edge or in the Cloud depending on the time criticality requested. Cybersecurity is obviously an important point to ensure a seamless manufacturing process.

Figure 12.5 shows the entire Industry 4.0 environment. The physical part gathers equipment such as machine tools or robots. To control them, the network between the Edge data center and the end devices must react to the millisecond. Until now, this was mainly the job of the field networks. However, they are complex because the commands pass through them in series, while the different actuators work in parallel. This is why there is a strong inclination toward 5G, which allows directional antennas to issue commands in parallel. Another possibility is to use embedded data centers at the machine tool level to have even much lower reaction times. However, these embedded data centers must be synchronized to reach the different actuators in a synchronous way. The direct mode going directly from an embedded data center to another embedded data center is a good solution.

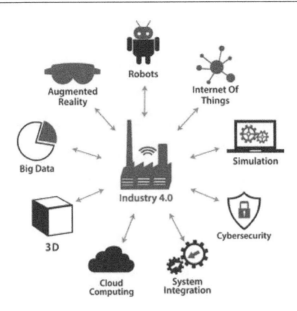

Figure 12.4. *The main elements of Industry 4.0. For a color version of this figure, see www.iste.co.uk/haddadou/edge.zip*

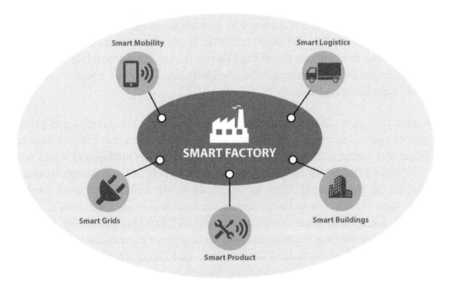

Figure 12.5. *The Industry 4.0 environment. For a color version of this figure, see www.iste.co.uk/haddadou/edge.zip*

We have seen that three levels of data centers are used. The first is represented by the Cloud to perform significant tasks such as managing Big Data or sophisticated machine learning. The second is the Edge with Fog or MEC depending on the type of network used, and finally the embedded Edge in the end machine or extremely close to it. We can see these three levels in Figure 12.6.

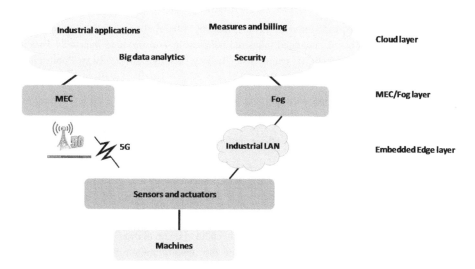

Figure 12.6. *The three major tiers of data centers used by Industry 4.0. For a color version of this figure, see www.iste.co.uk/haddadou/edge.zip*

12.4. Conclusion

Industry 4.0 represents a new generation of applications that require both real-time and reliability properties from the communication environment. Industry 4.0 relies on intelligent controls capable of reacting in millisecond timeframes. 5G, with its new mission-critical and ultra-high reliability properties, represents the core network for this new world. This leads to an architecture in which sensors, actuators and machines are connected directly to an antenna that can be inside the enterprise with private 5G or outside the enterprise using public 5G. However, embedded Edge solutions are also capable of addressing the challenges posed by Industry 4.0 through the presence of embedded data centers in machines located on the endpoint of the Edge located in the machine tools themselves. We find the same dichotomy of the automation world in which we either outsource to a 5G operator or we keep control in the company by placing the necessary equipment to keep control of the applications.

12.5. References

Ahmed, E. and Rehmani, M.H. (2017). Mobile Edge Computing: Opportunities, solutions, and challenges. *Future Generation Computer Systems*, 70, 59–63.

Al-Emran, M., Mezhuyev, V., Kamaludin, A. (2018). Technology acceptance model in m-learning context: A systematic review. *Computers & Education*, 125, 389–412.

Bellandi, M., De Propris, L., Santini, E. (2019). Industry 4.0+ challenges to local productive systems and place based integrated industrial policies. In *Transforming Industrial Policy for the Digital Age*, Bianchi, P., Durán, C.R., Labory, S. (eds). Edward Elgar Publishing, Cheltenham.

Bibby, L. and Dehe, B. (2018). Defining and assessing Industry 4.0 maturity levels-case of the defence sector. *Production Planning & Control*, 29(12), 1030–1043.

Butt, J. (2020). A strategic roadmap for the manufacturing industry to implement Industry 4.0. *Designs*, 4(2), 11.

Choe, M.J. and Noh, G.Y. (2018). Combined model of technology acceptance and innovation diffusion theory for adoption of smartwatch. *International Journal of Contents*, 14(3), 32–38.

Dean, M. and Spoehr, J. (2018). The fourth industrial revolution and the future of manufacturing work in Australia: Challenges and opportunities. *Labour & Industry: A Journal of the Social and Economic Relations of Work*, 28(3), 166–181.

Fatorachian, H. and Kazemi, H. (2018). A critical investigation of Industry 4.0 in manufacturing: Theoretical operationalisation framework. *Production Planning & Control*, 29(8), 633–644.

Ghobakhloo, M. and Fathi, M. (2019). Corporate survival in Industry 4.0 era: The enabling role of lean-digitized manufacturing. *Journal of Manufacturing Technology Management*, 31, 1–30.

Luthra, S. and Mangla, S.K. (2018). Evaluating challenges to Industry 4.0 initiatives for supply chain sustainability in emerging economies. *Process Safety and Environmental Protection*, 117, 168–179.

Machado, C.G., Winroth, M.P., Ribeiro da Silva, E.H.D. (2020). Sustainable manufacturing in Industry 4.0: An emerging research agenda. *International Journal of Production Research*, 58(5), 1462–1484.

Masood, T. and Sonntag, P. (2020). Industry 4.0: Adoption challenges and benefits for SMEs. *Computers in Industry*, 121, 103261.

Mohamed, M. (2018). Challenges and benefits of Industry 4.0: An overview. *International Journal of Supply and Operations Management*, 5(3), 256–265.

Moktadir, M.A., Ali, S.M., Kusi-Sarpong, S., Shaikh, M.A.A. (2018). Assessing challenges for implementing Industry 4.0: Implications for process safety and environmental protection. *Process Safety and Environmental Protection*, 117, 730–741.

Pace, P., Aloi, G., Gravina, R., Caliciuri, G., Fortino, G., Liotta, A. (2019). An edge-based architecture to support efficient applications for healthcare industry 4.0. *IEEE Transactions on Industrial Informatics*, 15(1), 481–489.

Sharma, M., Kamble, S., Mani, V., Sehrawat, R., Belhadi, A., Sharma, V. (2021). Industry 4.0 adoption for sustainability in multi-tier manufacturing supply chain in emerging economies. *Journal of Cleaner Production*, 281, 125013.

Strandhagen, J.W., Alfnes, E., Strandhagen, J.O., Vallandingham, L.R. (2017). The fit of Industry 4.0 applications in manufacturing logistics: A multiple case study. *Advances in Manufacturing*, 5(4), 344–358.

13

AI for Cloud and Edge Networking

Artificial intelligence (AI) is appearing more and more as one of the functionalities one can expect from a Cloud. The advantage of this solution is being able to simply add a virtual machine in the Cloud next to the virtual machines of application service, virtual infrastructure service and the digital infrastructure itself. This is the way to accelerate the automation of networks, to optimize the radio of a wireless communication or to diagnose a system failure. In this chapter, we will look at the introduction of AI to optimize the operation of the Edge and Cloud Networking.

Where do we find uses for AI in Edge and Cloud Networking? First, in network management systems, one of the clear goals for the end of this decade is to achieve global automation of all processes that take place in a network. There is also a lot of AI in cybersecurity, which is becoming more and more complex with digitization. Terminal equipment is also full of AI to improve photos, to optimize the use of resources, to manage user access and finally, without being at all exhaustive, to optimize its applications.

There are more and more smart spaces or smart bubbles such as the smart city, Industry 4.0 or the tactile Internet.

13.1. The knowledge plane

The knowledge plane is one of the characteristics of Cloud Networking. In order to obtain an automatic steering of the network, it is necessary to have enough knowledge, that is, information with its context, to be able to obtain an automatic steering. It is therefore necessary to gather this knowledge either in a central point that has the steering, or in each point of the network to have a distributed control that is more difficult to attack. In this case, all the nodes must have a global view of the system so that convergence of all the nodes is possible.

The knowledge plane is centralized in the 2000–2025 generation because it provides a simple solution to gather all the knowledge without having to send it to all the nodes in the network. All the knowledge goes back to a central node, the controller, and all the decision intelligence is in this equipment, which is usually virtualized and runs in a data center.

Looking at the current research, the knowledge plane should go toward distribution to get a fully distributed network autopilot for the 2030s and specifically for 6G.

The knowledge plane and the controller that supports it are increasingly using Big Data and real-time analysis techniques, so the pilot can decide what actions to take. Big Data is characterized by the four Vs:

– V for volume: an annual increase of more than 50%.

– V for velocity: in addition to the rapid obsolescence of some of these real-time data, there is the need to integrate other data as quickly as possible in order to generate the most up-to-date information.

– V for variety: to the diversity of sources and formats (text, photo, video, sound, technical log), especially of unstructured data, is added a great variety of internal and external suppliers, objects or persons.

– V for value: value-creating data are the most interesting.

Data can be digitally formatted but also non-digital according to several categories. Unstructured data are presented in the form of:

– free text;

– images;

– sound (voice);

– video;

– logs of connected objects.

Figure 13.1 shows the differences between information management without Big Data and with Big Data.

AI for Cloud and Edge Networking   213

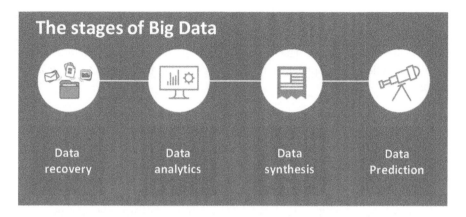

Figure 13.1. *Information management with and without Big Data. For a color version of this figure, see www.iste.co.uk/haddadou/edge.zip*

Figure 13.2 details how to manage and govern Big Data from the data sources, their storage so that they can be easily and efficiently exploited, the importance of metadata in their use and finally at the top of the scale the data that can be directly exploited with associated tools.

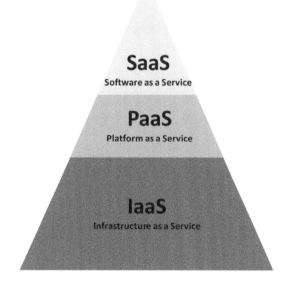

Figure 13.2. *Data governance. For a color version of this figure, see www.iste.co.uk/haddadou/edge.zip*

To conclude this section, it is clear that Big Data is one of the most widely used tools to drive, manage, control and optimize networks. Big Data in memory, that is, stored in the RAMs of the data center, allows for moving into real time. Most network and telecom equipment manufacturers use this technology to bring about intelligence.

13.2. Artificial intelligence and Software-Defined Networking

SDN is the basic technology for controlling SDN networks that form the backbone of Cloud Networking. The knowledge plane is implemented in the controller. In recent years, the idea has been to animate the controller with AI as shown in Figure 13.3. The knowledge ranges from the network via the SDN switches to the controller. The controller has a brain that can take actions based on the knowledge gathered from the network without stopping.

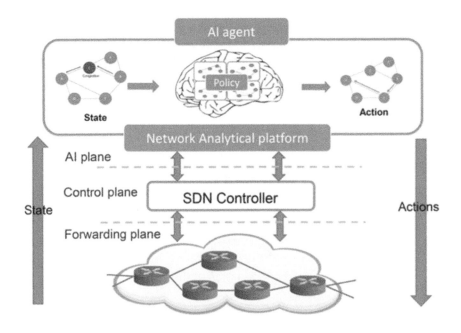

Figure 13.3. *Automatic and intelligent control of an SDN network. For a color version of this figure, see www.iste.co.uk/haddadou/edge.zip*

Various AI technologies are used, including Big Data and machine learning, always in a centralized manner. The controller makes decisions and applies them through the network nodes that receive the configurations to be implemented.

In a 5G environment where slicing applies, each virtual network uses AI-driven SDN. Each slice can have a different driving technology since the slices are independent of each other. One of the difficulties comes from the sharing of resources between the slices.

Studies in the early 2020s are leading to improvements in these techniques, sometimes not to improve them but to make them simpler and less energy intensive, which can result in a substantial gain in the end. Figure 13.4 symbolizes current improvements to simplify control and optimize temp loss due to too many configurations. To do this, we need to find an average configuration that is not completely optimal but near optimal. This technique is called IBN (Intend-Based Networking). We have already seen it in SDN architectures, especially at Cisco.

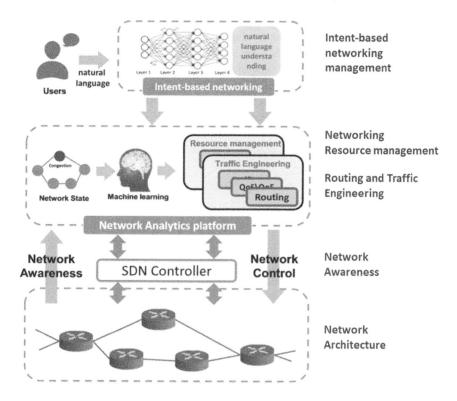

Figure 13.4. *Intent-based networking (IBN). For a color version of this figure, see www.iste.co.uk/haddadou/edge.zip*

In this technology, we need to find the network configuration that adapts to all user demands without having to reconfigure the SDN nodes over and over again. To do this, there are five steps:

– gathering the intentions of all network participants;

– making a working model that knows how to introduce the intentions;

– validating the result of the working model to prove that it is acceptable by the network and that the users' QoS are satisfied;

– having an autopilot system to automatically configure the nodes;

– deploying configurations.

Figure 13.5 outlines the steps to arrive at this solution. First, we need to recover all the intentions of the potential users of the network. It is also necessary to have a strong knowledge of the network infrastructure with all the elements influencing the performance and certainly the available security. Then, all this knowledge must be processed in the controller or in a virtual machine to obtain an optimal configuration with respect to all the intentions. A validation system must be exercised to verify that each client can count on an adequate quality of service and security. Finally, all the elements participating in the network must be configured. This solution enables obtaining a network that does not need to reconfigure itself continuously to adapt to each incoming flow. However, it seems that the network infrastructure must be a little more powerful than what is necessary in a very fine optimization, but the gains are not negligible because there is no more time loss for the numerous reconfigurations necessary in the case of instantaneous optimization.

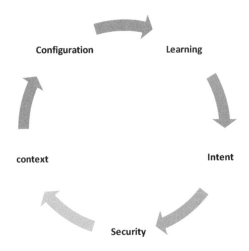

Figure 13.5. *Diagram of actions to be taken to achieve IBN. For a color version of this figure, see www.iste.co.uk/haddadou/edge.zip*

AI is particularly important in this first aspect of network configuration optimization. However, we must be careful that the high use of network resources to optimize it does not result in becoming counterproductive and achieving lower performance. A fair compromise must be found between optimization and the resources used to carry out the optimization.

13.3. AI and Cloud Networking management

AI is a state-of-the-art technique for performing diagnostics when anomalies occur in the network. Similarly, the automatic steering of a new generation network requires a high level of intelligence as soon as the network is large in number of nodes. The objective of this section is not to look at all the possible and imaginable advances of AI but to give an example to understand what can be done with it. For that, we will explain a use case which is the detection of failures for any reason but especially the prediction of these failures before they occur.

This example comes from the company PacketAI, which offers a product capable of detecting that an outage is going to occur, diagnosing the reasons that will lead to the outage and predicting the approximate time when it will occur. The idea is to use all of the management data, to bring them into a processing machine which is in fact a virtual machine in a Cloud because the mass of information is so large. In the case of management, real time is not necessary as opposed to control, and the equipment that supports the virtual machine can be located relatively far from the network although the cost of transport can become prohibitive to gather all the necessary data and knowledge. Networking is one of the most information-gathering environments. Each piece of equipment produces a large amount of information that is often discarded because it is usable by a human. If we form Big Data with all the information coming from the network infrastructure, we need a large storage system, and if we want to access low response times, we also need a lot of RAM. This Big Data is analyzed to draw correlations that humans cannot detect in the huge amount of data. These correlations are in fact correlated to failures, but this is only detected through AI such as machine learning or deep learning, which is also capable of estimating the time of failure. To achieve this, it is necessary to process colossal masses of data that must be cleaned and sorted in such a way that the cost of detection is not negligible in terms of energy spent and machine power.

In the same vein, network autopilots are being developed in open-source form. We saw in Chapter 4 the ONAP project (Open Network Automation Project) which seems by far the best placed to become the standard. ONAP has several internal controllers for configuring, managing and controlling network equipment and applications. In the same way as before, it is necessary to collect a maximum of data on the various points concerning the controllers, and with machine learning the Edge and Cloud Networking can be driven automatically. One of the problems is the size of the network. The tests carried out show that the network functions very well when it has a limited number of nodes. But as soon as the size exceeds a dozen nodes, the control becomes more approximate. It is therefore necessary to move toward a division into clusters with an overlay network of automatic drivers. It is also necessary to progress on distributed pilots which form a burst system on several nodes, to obtain an automatic piloting of large networks.

One of the directions for this demand for power and intelligent processes is to develop data centers specialized in AI. This is one of the goals for many companies providing cloud services.

13.4. AI through digital twins

Digital twins are a new paradigm that is rapidly rising in both the service industry and the industrial world. It is about copying a physical machine in a perfect way, usually in software form. The digital twin behaves exactly the same way as the physical machine. This property is illustrated in Figure 13.6.

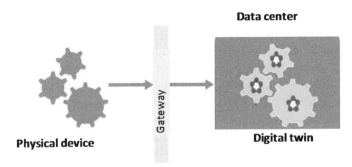

Figure 13.6. *A digital twin. For a color version of this figure, see www.iste.co.uk/haddadou/edge.zip*

Many digital twins are deployed in the world to control the associated physical object. Examples are numerous: aircraft engines at Airbus and Boeing, vehicles at Tesla, smart cities, etc. The digital twin can also be used to protect the object to which it is attached. Indeed, the only way to access the object may be through its twin, which may be strongly defended, whereas the physical object has no defense. But the most often cited objective is to allow the twin to get ahead of the physical object to detect failures or abnormal behavior.

The digital twin can be a simulation model, a logical virtual machine or a physical virtual machine. The next step, once the twin is perfectly defined, is to transport the data from the physical object to the logical object in real time if possible and with security so that the decisions made by the digital twin are not distorted. The next step is to execute the digital twin to analyze and understand the lifecycle of the physical object. Then, it is possible to monitor, in real time if necessary, the physical object by analyzing its behavior to understand the lifecycle. The final objective is prediction to detect problems and try to repair what is necessary to avoid failure. Finally, the feedback modifies the parameters and repair so that the physical object can continue its lifecycle quietly. Figure 13.7 shows what has just been decided. The time axis shows the execution of the physical object. The speed of execution of the object is indicated by the oblique line indicating the state of the digital twin, which in this case is at a speed higher than that of the physical object. The detection of an anomaly is done in advance compared to the physical machine, which leaves a slight lapse of time to realize the repair.

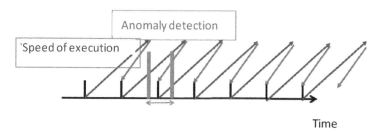

Figure 13.7. *Behavior of a digital twin. For a color version of this figure, see www.iste.co.uk/haddadou/edge.zip*

To summarize the behaviors of a digital twin, the three steps that must occur are described in Figure 13.8.

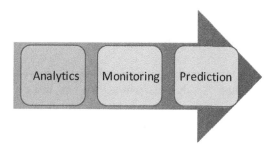

Figure 13.8. *The three stages of operation of a digital twin. For a color version of this figure, see www.iste.co.uk/haddadou/edge.zip*

While digital twins, in the early 2020s, are more often associated with somewhat high-end objects and therefore require significant power, they are becoming smaller and smaller to control objects of a few dozen euros. They become the guardians of the physical object, and they control their functioning by AI coming from virtual machines implanted in the digital twin. Figure 13.9 describes a set of digital twins associated with objects and themselves using AI, security or automation processes.

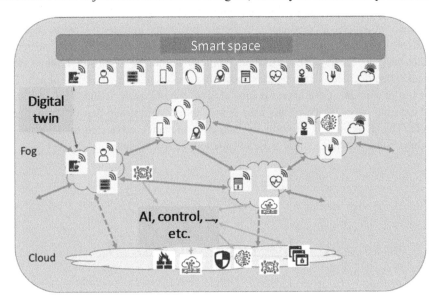

Figure 13.9. *Digital twins associated with objects. For a color version of this figure, see www.iste.co.uk/haddadou/edge.zip*

Digital twins are growing rapidly and are expected to become commonplace in the coming years to control millions or even billions of objects. They will take a key role in controlling large systems such as 6G technology, as visualized by the Japanese operator NTT-Docomo, which presents the solution described in Figure 13.10 to manage and control this complex environment.

In this figure, we distinguish the physical space in which the whole 6G environment is positioned. To this real world, we can make it correspond to a cyberspace which is fed by all the knowledge coming from the sensors and the equipment collecting the data of the physical world. The cyberspace with its continuous supply of knowledge makes control decisions that are sent to the physical world via actuators. Of course, cyberspace must be able to operate in real time and even be able to outpace the physical world. This requires phenomenal data center power, which is usually found in large data centers with more than a million servers.

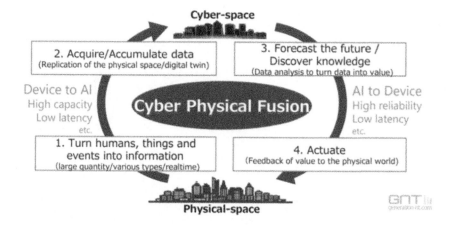

Figure 13.10. *NTT Docomo's view of 6G control. For a color version of this figure, see www.iste.co.uk/haddadou/edge.zip*

13.5. Conclusion

As we have seen, digital twins represent one of the great ways to manage, control or secure objects from the small to the very large. In the example of NTT-Docomo's vision, the cyber-space appears centralized as a huge data center, but it may be distributed into a greater or lesser number of virtual machines. If they are many very small virtual machines, there is significant research to be done on showing the convergence of the system control.

6G is also a source for looking at what is intelligence and where to position it. Figure 13.11 describes the vision of Nokia, which is also that of the large European project for 6G called Hexa-X. Three worlds form the whole system: the digital world, the physical world and the human world. These worlds generate the 6G system which is controlled in real time with digital twins and biological intelligence.

Biological intelligence can be defined as the intelligence that comes from humans or plants. The characteristics are love, empathy, sensitivity, etc. It is human instinct. This intelligence has been under research for at least 10 years, and significant progress is beginning to be made.

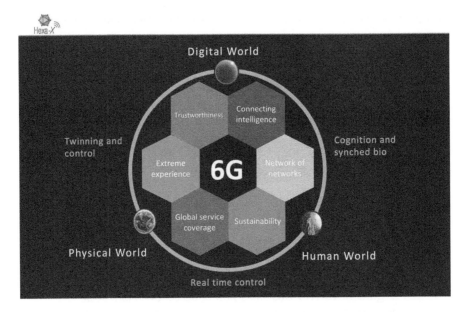

Figure 13.11. *The structure of 6G by the Hexa-X project led by Nokia. For a color version of this figure, see www.iste.co.uk/haddadou/edge.zip*

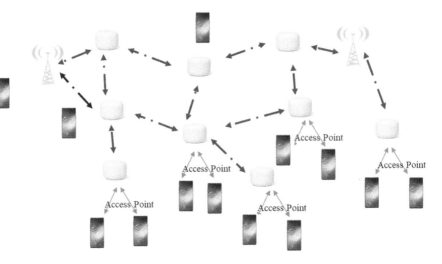

Figure 13.12. *A possible infrastructure for 6G. For a color version of this figure, see www.iste.co.uk/haddadou/edge.zip*

Figure 13.12 describes a possible infrastructure for 6G that is being pushed by members of GAFAM, in particular Apple and Google. The data centers are in the user's pocket or on the desktop. These data centers are interconnected by a mesh network. The virtual machines are distributed in the femto data centers.

13.6. References

Afazov, S. and Scrimieri, D. (2020). Chatter model for enabling a digital twin in machining. *The International Journal of Advanced Manufacturing Technology*, 110(9), 2439–2444.

Aheleroff, S., Xu, X., Zhong, R.Y., Lu, Y. (2021). Digital twin as a service (DTaaS) in industry 4.0: An architecture reference model. *Advanced Engineering Informatics*, 47, 101225.

Aivaliotis, P., Georgoulias, K., Chryssolouris, G. (2019). The use of digital twin for predictive maintenance in manufacturing. *International Journal of Computer-Integrated Manufacturing*, 32(11), 1067–1080.

Alexopoulos, K., Nikolakis, N., Chryssolouris, G. (2020). Digital twin-driven supervised machine learning for the development of artificial intelligence applications in manufacturing. *International Journal of Computer-Integrated Manufacturing*, 33(5), 429–439.

Anwer, N., Liu, A., Wang, L., Nee, A.Y., Li, L., Zhang, M. (2021). Digital twin towards smart manufacturing and industry 4.0. *Journal of Manufacturing Systems*, 58, 1–2.

Barricelli, B.R., Casiraghi, E., Fogli, D. (2019). A survey on digital twin: Definitions, characteristics, applications, and design implications. *IEEE Access*, 7, 167653–167671.

Cao, X., Zhao, G., Xiao, W. (2020). Digital twin-oriented realtime cutting simulation for intelligent computer numerical control machining. *Proceedings of the Institution of Mechanical Engineers, Part B: Journal of Engineering Manufacture*, 236(1–2), 5–15.

Cheng, J., Zhang, H., Tao, F., Juang, C.F. (2020). DT-II: Digital twin enhanced Industrial Internet reference framework towards smart manufacturing. *Robotics and Computer-Integrated Manufacturing*, 62, 101881.

Cruz-Martinez, G.M. and Z-Avilez, L.A. (2020). Design methodology for rehabilitation robots: Application in an exoskeleton for upper limb rehabilitation. *Applied Sciences*, 10(16), 5459.

Dhiman, H. and Rocker, C. (2021). Middleware for providing activity-driven assistance in cyber-physical production systems. *Journal of Computational Design and Engineering*, 8(1), 428–451.

Ding, K., Chan, F.T., Zhang, X., Zhou, G., Zhang, F. (2019). Defining a digital twin-based cyber-physical production system for autonomous manufacturing in smart shop floors. *International Journal of Production Research*, 57(20), 6315–6334.

Fera, M., Greco, A., Caterino, M., Gerbino, S., Caputo, F., Macchiaroli, F., D'Amato, E. (2020). Towards digital twin implementation for assessing production line performance and balancing. *Sensors*, 20(1), 97.

Glatt, M., Sinnwell, C., Yi, L., Donohoe, S., Ravani, B., Aurich, J.C. (2021). Modeling and implementation of a digital twin of material flows based on physics simulation. *Journal of Manufacturing Systems*, 58, 231–245.

Guo, H., Chen, M., Mohamed, K., Qu, T., Wang, S., Li, J. (2021). A digital twin-based flexible cellular manufacturing for optimization of air conditioner line. *Journal of Manufacturing Systems*, 58, 65–78.

Han, S. (2020). A review of smart manufacturing reference models based on the skeleton meta-model. *Journal of Computational Design and Engineering*, 7(3), 323–336.

Han, Z., Li, Y., Yang, M., Yuan, Q., Ba, L., Xu, E. (2020). Digital twin-driven 3D visualization monitoring and traceability system for general parts in continuous casting machine. *Journal of Advanced Mechanical Design, Systems, and Manufacturing*, 14(7), AMDSM0100-JAMDSM0100.

Hwang, D. and Do Noh, S. (2028). A study on the integrations of products and manufacturing engineering using sensors and IoT. In *IFIP International Conference on Advances in Production Management Systems*, pp. 370–377. Springer.

Jeon, B., Yoon, J.S., Um, J., Suh, S.H. (2020). The architecture development of Industry 4.0 compliant smart machine tool system (SMTS). *Journal of Intelligent Manufacturing*, 31(8), 1837–1859.

Kang, S., Chun, I., Kim, H.S. (2019). Design and implementation of runtime verification framework for cyber-physical production systems. *Journal of Engineering*, 2019(2875236), 11.

Kannan, K. and Arunachalam, N. (2019). A digital twin for grinding wheel: An information sharing platform for sustainable grinding process. *Journal of Manufacturing Science and Engineering*, 141(2), 021015.

Kim, H.C., Son, Y.H., Bae, J., Noh, S.D. (2021). A study on the digital twin visualization method for smart factory of dye processing industry. *Journal of the Korean Institute of Industrial Engineers*, 47(1), 77–91.

Lee, I.D., Lee, I., Ha, S. (2021). 3D reconstruction of as-built model of plant piping system from point clouds and port information. *Journal of Computational Design and Engineering*, 8(1), 195–209.

Leng, J., Yan, D., Liu, Q., Zhang, H., Zhao, G., Wei, L., Zhang, D., Yu, A., Chen, X. (2019). Digital twin-driven joint optimisation of packing and storage assignment in large-scale automated high-rise warehouse product-service system. *International Journal of Computer-Integrated Manufacturing*, 34, 1–18.

Leng, J., Liu, Q., Ye, S., Jing, J., Wang, Y., Zhang, C., Ding, Z., Chen, X. (2020). Digital twin-driven rapid reconfiguration of the automated manufacturing system via an open architecture model. *Robotics and Computer-Integrated Manufacturing*, 63, 101895.

Lim, K.Y.H., Zheng, P., Chen, C.H. (2019). A state-of-the art survey of digital twin: Techniques, engineering product lifecycle management and business innovation perspectives. *Journal of Intelligent Manufacturing*, 31, 1–25.

Lim, K.Y.H., Zheng, P., Chen, C.H., Huan, L. (2020). A digital twin-enhanced system for engineering product family design and optimization. *Journal of Manufacturing Systems*, 57, 82–93.

Liu, J., Zhou, H., Liu, X., Tian, G., Wu, M., Cao, L., Wang, W. (2019a). Dynamic evaluation method of machining process planning based on digital twin. *IEEE Access*, 7, 19312–19323.

Liu, Q., Zhang, H., Leng, J., Chen, X. (2019b). Digital twin-driven rapid individualised designing of automated flow-shop manufacturing system. *International Journal of Production Research*, 57(12), 3903–3919.

Liu, S., Bao, J., Lu, Y., Li, J., Lu, S., Su, X. (2021). Digital twin modeling method based on biomimicry for machining aerospace components. *Journal of Manufacturing Systems*, 58, 180–195.

Liu, C., Le Roux, L., Korner, C., Tabaste, O., Lacan, F., Bigot, S. (2022). Digital twin-enabled collaborative data management for metal additive manufacturing systems. *Journal of Manufacturing Systems*, 62, 857–874.

Lu, Y., Liu, C., Kevin, I., Wang, K., Huang, H., Xu, X. (2020). Digital twin-driven smart manufacturing: Connotation, reference model, applications and research issues. *Robotics and Computer-Integrated Manufacturing*, 61, 101837.

Luo, W., Hu, T., Ye, Y., Zhang, C., Wei, Y. (2020). A hybrid predictive maintenance approach for CNC machine tool driven by digital twin. *Robotics and Computer-Integrated Manufacturing*, 65, 101974.

Ma, J., Chen, H., Zhang, Y., Guo, H., Ren, Y., Mo, R., Liu, L. (2020). A digital twin-driven production management system for production workshop. *The International Journal of Advanced Manufacturing Technology*, 110(5), 1385–1397.

Mi, S., Feng, Y., Zheng, H., Wang, Y., Gao, Y., Tan, J. (2021). Prediction maintenance integrated decision-making approach supported by digital twin-driven cooperative awareness and interconnection framework. *Journal of Manufacturing Systems*, 58, 329–345.

Oyekan, J., Farnsworth, M., Hutabarat, W., Miller, D., Tiwari, A. (2020). Applying a 6 DoF robotic arm and digital twin to automate fan-blade reconditioning for aerospace maintenance, repair, and overhaul. *Sensors*, 20(16), 4637.

Papetti, A., Gregori, F., Pandolfi, M., Peruzzini, M., Germani, M. (2020). A method to improve workers' well-being toward human-centered connected factories. *Journal of Computational Design and Engineering*, 7(5), 630–643.

Park, J., Samarakoon, S., Bennis, G., Debbah, M. (2019a). Wireless Network Intelligence at the Edge. *IEEE 2019*, 107, 2204–2239.

Park, K.T., Im, S.J., Kang, Y.S., Noh, S.D., Kang, Y.T., Yang, S.G. (2019b). Service-oriented platform for smart operation of dyeing and finishing industry. *International Journal of Computer-Integrated Manufacturing*, 32(3), 307–326.

Park, K.T., Lee, D., Do Noh, S. (2020a). Operation procedures of a work-center-level digital twin for sustainable and smart manufacturing. *International Journal of Precision Engineering and Manufacturing-Green Technology*, 7(3), 791–814.

Park, Y., Woo, J., Choi, S. (2020b). A cloud-based digital twin manufacturing system based on an interoperable data schema for smart manufacturing. *International Journal of Computer-Integrated Manufacturing*, 33(12), 1259–1276.

Pires, F., Cachada, A., Barbosa, J., Moreira, A.P., Leitao, P. (2019). Digital twin in Industry 4.0: Technologies, applications and challenges. *IEEE 17th International Conference on Industrial Informatics (INDIN)*, 1, 721–726.

Polini, W. and Corrado, A. (2020). Digital twin of composite assembly manufacturing process. *International Journal of Production Research*, 58(17), 5238–5252.

Qi, Q. and Tao, F. (2018). Digital twin and Big Data towards smart manufacturing and Industry 4.0: 360 degree comparison. *IEEE Access*, 6, 3585–3593.

Rasheed, A., San, O., Kvamsdal, T. (2020). Digital twin: Values, challenges and enablers from a modeling perspective. *IEEE Access*, 8, 21980–22012.

Rauch, E. and Vickery, A.R. (2020). Systematic analysis of needs and requirements for the design of smart manufacturing systems in SMEs. *Journal of Computational Design and Engineering*, 7(2), 129–144.

Roy, R.B., Mishra, D., Pal, S.K., Chakravarty, T., Panda, S., Chandra, M.G., Pal, A., Misra, P., Chakravarty, D., Misra, S. (2020). Digital twin: Current scenario and a case study on a manufacturing process. *The International Journal of Advanced Manufacturing Technology*, 107(9), 3691–3714.

Scafa, M., Marconi, M., Germani, M. (2020). A critical review of symbiosis approaches in the context of Industry 4.0. *Journal of Computational Design and Engineering*, 7(3), 269–278.

Schroeder, G.N., Steinmetz, C., Rodrigues, R.N., Henriques, R.V.B., Rettberg, A., Pereira, C.E. (2020). A methodology for digital twin modeling and deployment for Industry 4.0. *Proceedings of the IEEE*, 109, 556–567.

Semeraro, C., Lezoche, M., Panetto, H., Dassisti, M. (2021). Digital twin paradigm: A systematic literature review. *Computers in Industry*, 130, 103469.

Shikata, H., Yamashita, T., Arai, K., Nakano, T., Hatanaka, K., Fujikawa, H. (2019). Digital twin environment to integrate vehicle simulation and physical verification. *SEI Technical Review*, 88, 18–21.

Son, Y.H., Park, K.T., Lee, D., Jeon, S.W., Noh, S.D. (2021). Digital twin–based cyber-physical system for automotive body production lines. *The International Journal of Advanced Manufacturing Technology*, 115, 291–310.

Stoumpos, S., Theotokatos, G., Mavrelos, C., Boulougouris, E. (2020). Towards marine dual fuel engines digital twins – Integrated modeling of thermodynamic processes and control system functions. *Journal of Marine Science and Engineering*, 8(3), 200.

Sun, X., Bao, J., Li, J., Zhang, Y., Liu, S., Zhou, B. (2020). A digital twin-driven approach for the assembly-commissioning of high precision products. *Robotics and Computer-Integrated Manufacturing*, 61, 101839.

Tan, Y., Yang, W., Yoshida, K., Takakuwa, S. (2019). Application of IoT-aided simulation to manufacturing systems in cyberphysical system. *Machines*, 7(1), 2.

Tong, X., Liu, Q., Pi, S., Xiao, Y. (2019). Real-time machining data application and service based on IMT digital twin. *Journal of Intelligent Manufacturing*, 31, 1–20.

Wang, K.J., Lee, Y.H., Angelica, S. (2020). Digital twin design for real-time monitoring: A case study of die cutting machine. *International Journal of Production Research*, 59, 1–15.

Xia, K., Sacco, C., Kirkpatrick, M., Saidy, C., Nguyen, L., Kircaliali, A., Harik, R. (2021). A digital twin to train deep reinforcement learning agent for smart manufacturing plants: Environment, interfaces and intelligence. *Journal of Manufacturing Systems*, 58, 210–230.

Xu, X. (2012). From cloud computing to cloud manufacturing. *Robotics and Computer-Integrated Manufacturing*, 28(1), 75–86.

Xu, W., Cui, J., Li, L., Yao, B., Tian, S., Zhou, Z. (2020). Digital twin-based industrial cloud robotics: Framework, control approach and implementation. *Journal of Manufacturing Systems*, 58, 196–209.

Yi, Y., Yan, Y., Liu, X., Ni, Z., Feng, J., Liu, J. (2021). Digital twin-based smart assembly process design and application framework for complex products and its case study. *Journal of Manufacturing Systems*, 58, 94–107.

Zhang, M., Tao, F., Nee, A.Y.C. (2021). Digital twin enhanced dynamic job-shop scheduling. *Journal of Manufacturing Systems*, 58, 146–156.

Zhao, P., Liu, J., Jing, X., Tang, M., Sheng, S., Zhou, H., Liu, X. (2020). The modeling and using strategy for the digital twin in process planning. *IEEE Access*, 8, 41229–41245.

Zheng, Y., Chen, L., Lu, X., Se, Y., Cheng, H. (2020). Digital twin for geometric feature online inspection system of car body in-white. *International Journal of Computer Integrated Manufacturing*, 34, 1–12.

Zhong, D., Lin, P., Lyu, Z., Rong, Y., Huang, G.Q. (2020). Digital twin-enabled graduation intelligent manufacturing system for fixed-position assembly islands. *Robotics and Computer-Integrated Manufacturing*, 63, 101917.

Zhou, G., Zhang, C., Li, Z., Ding, K., Wang, C. (2020). Knowledge driven digital twin manufacturing cell towards intelligent manufacturing. *International Journal of Production Research*, 58(4), 1034–1051.

Zhuang, C., Gong, J., Liu, J. (2021). Digital twin-based assembly data management and process traceability for complex products. *Journal of Manufacturing Systems*, 58, 118–131.

14

Cloud and Edge Networking Security

Edge and Cloud networking security is complex. Many parameters are involved, both at the physical and logical levels. In this chapter, we will describe a number of solutions deployed in companies but which are sometimes limited by the complexity coming from both the network and the system side. We will start by examining Cloud Security which is a new paradigm whose objective is not to bring security to Clouds but to propose virtual machines whose objective is to make the rest of the world secure. Next, we will look at SIM cards and their evolution in the context of Cloud Networking. Finally, we will focus on several solutions to secure the Cloud.

14.1. The Security Cloud

The Security Cloud is a new paradigm that should spread rapidly in the 2020s, that of simplifying security processes by gathering them all in virtual machines that are housed in an adapted data center. We find firewalls, which today form the majority of virtual machines, but also authentication or identification servers, attack or DDOS detection processes, various and varied filters, in particular DPI (Deep Packet Inspection), which detects applications transported on a network by examining the flow of binary elements. There are also HSMs (Hardware Security Modules) for companies and individuals. The HSM, which we will review later, is a form of equipment to protect cryptographic actions.

If we start with firewalls, which are the most widespread case, one of the ideas is to develop firewalls by application by going into great detail about the security of a particular application. A DPI recognizes the application and sends the corresponding stream to the virtualized firewall, which has the necessary power thanks to the Cloud to dive into the details of the session to detect possible misuse.

14.2. SIM-based security

Figure 14.1 describes the elements of the first solution we have chosen, in which a locally secure element must be possessed. This secure element can be a SIM card from a telecommunications operator in a more or less miniaturized form. The secure element comes either from the telecom operator, or it is soldered on the processor of the machine to be defended, or it belongs to the equipment manufacturer who made the terminal. Two other solutions are to be noted but not widely used: a secure element embedded in an SD card and inserted in the reader of the mobile terminal and an external secure element communicating with the mobile by an NFC (Near Field Communication) interface. These different solutions have been strongly developed through mobile networks to secure terminals. The equipment manufacturers who have been able to make a sufficient place for themselves on the market of smartphones, tablets and portable equipment have chosen this solution. The secure element, as shown in Figure 14.1, is a smart card or an embedded secure element, for example, from the company NXP, as in Apple smartphones.

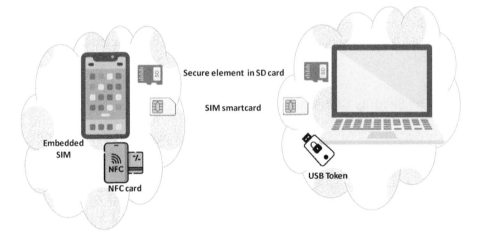

Figure 14.1. *Security elements. For a color version of this figure, see www.iste.co.uk/haddadou/edge.zip*

This solution is simple to set up; it has the advantage of being able to carry out the communications locally but also the disadvantages of having a strong limitation of the amount of software that can be embedded in a secured element and the difficulty of configuring the secured element when adding a new service.

The second major solution is the eSIM (embedded SIM), which works with profiles that are stored remotely. This solution can be seen as a virtual card solution: instead of being directly in the secure element, the profile is located remotely. To reach it, the eSIM has to open a secure channel to the virtual card that takes the initiative to replace the smartphone on which the eSIM is located. The immediate advantage is obviously having a very large number of virtual SIM cards attached to the same eSIM. The virtual card can be physical with a specific SIM card or a hardware device such as an HSM. In the first case, the physical SIM card is located for example at the service manager. In the second case, the HSM is located in a secure service provider.

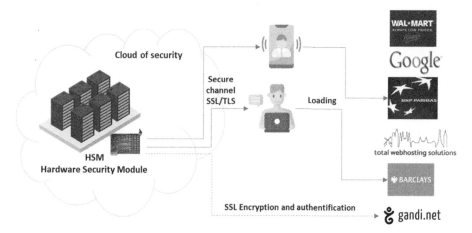

Figure 14.2. *How an eSIM works. For a color version of this figure, see www.iste.co.uk/haddadou/edge.zip*

Figure 14.2 shows the communications related to an eSIM. The smartphone or personal computer has an eSIM. This eSIM opens a secure channel with the virtual SIM card symbolized in the figure by either an electronic board with a few dozen physical SIM cards or an HSM. These elements can be located in a security Cloud, as we will see later. Once the secure channel is opened, the terminal has a secure element that is located remotely. The virtual SIM card takes care of the communication with the merchant to make a purchase. Once the purchase is completed, for example, digital music, the purchase is transported to the smartphone or laptop.

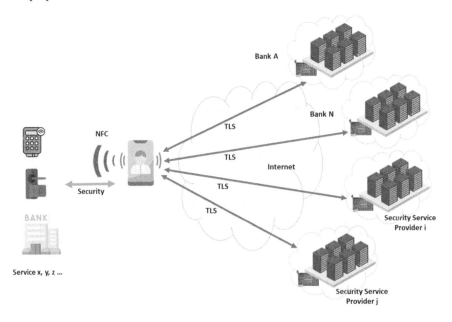

Figure 14.3. *The profiles linked to the eSIM card. For a color version of this figure, see www.iste.co.uk/haddadou/edge.zip*

Specifically, communication takes place between the mobile terminal and the virtual secure element, which is located in a security Cloud, for example. This communication takes place after opening a secure tunnel, once the opening is done with the SSL/TLS protocol. As soon as this communication is established, the secure element establishes a relationship with the merchant, always through a secure channel that opens between the secure element and the merchant site. Purchases can be made in complete security. Once the purchase is made, the task is passed back to the mobile terminal, which can retrieve the music or video purchased or, indirectly, any type of purchase. The secure element can be in a secure element Cloud, but also in an HSM that plays the same role, provided that the customer has confidence in this equipment. This is why secure element Clouds are still the most likely to develop because the secure elements, which can be smartcards, can be owned by the user, and the user therefore has a high level of trust in this system.

A first advantage of this eSIM card solution is the ability to assign as many profiles as needed to a single user. The customer can have several banks, several key managers and several security service providers. A security service provider can enter this market very simply by setting up its own secure element server or by positioning them at a security Cloud provider. This property is illustrated in Figure 14.3.

14.3. Blockchain and Cloud

Blockchain is a tool for certification and traceability. This technology can be used particularly in network environments. This technology was initially developed for cryptocurrencies such as bitcoin but has many other applications in networks, for example, to certify the identity of virtual machines to avoid usurpations of virtual machines.

Blockchain is a database technology that uses an immutable register. Transactions are recorded in this register and once they are recorded in the register, they cannot be modified. Anyone can consult this register and all the exchanges, present and past, without a trusted third party. It works directly in user-to-user mode. Being completely distributed, the system cannot be tampered with, and the verification of transactions can be done in the nodes that manage the chain, called miners.

The transactions that are to be encapsulated in the chain are collected in blocks, the last block being correlated to the previous one by the hash value of the previous block. All transactions in a new block are validated by the miners. If the block is valid, it is time stamped and permanently added to the chain. The transactions it contains are then visible throughout the network and can easily be verified by anyone who wants to. If the calculations are complex to perform (they are hash calculations), their verification is very simple. Once added to the chain, a block cannot be modified or deleted, which guarantees the authenticity and security of the whole.

As shown in Figure 14.4, each block in the chain consists of several transactions, a checksum (hash), used as an identifier, a checksum from the previous block (except for the first block in the chain, called the genesis block), and finally a measure of the amount of work that went into producing the block. The work is defined by the consensus method used within the chain, such as proof of work or proof of participation.

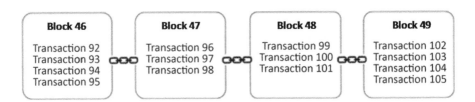

Figure 14.4. *How a blockchain works. For a color version of this figure, see www.iste.co.uk/haddadou/edge.zip*

The blockchain is a system based on trust. The system arrives, after more or less complex calculations, at a consensus that can no longer be questioned. The method to reach this consensus can be the proof of work or the proof of stake. The first method uses complex work to prove that the miner has succeeded in finding the right hash value of the block; the second takes into account the value of the assets. Indeed, the more a node is involved in the process, the less interest it has in the system stopping.

The protocol associated with the proof of work uses a cryptographic calculation requiring a significant amount of computing power, provided by the miners. This process cannot be broken, except in times of several thousand years. The miners are entities whose role is to feed the network with computing power, in order to allow the updating of the decentralized ledger. For this update, miners must be able to find a fairly complex hash. The first one to do this became the leader who managed the new block and won a reward of 6.25 bitcoins in 2022. It took about 15 min on extremely powerful machines.

A competition exists between miners to become the leader and take over the transactions of a block, allowing the power available on the network to grow. Anyone can use their computing power to mine, but the more miners there are, the harder it is to achieve consensus resolution. Thus, the protocol becomes almost unbreakable as soon as competition is strong between the nodes of the network, that is, no group of miners becomes the majority.

14.4. Cloud Networking security

Cloud security can be placed in the cyber security category but dedicated to securing Cloud networking systems. This security includes the protection of private data in the Edge and Cloud infrastructure, but also at the level of applications and processing platforms. Securing these systems involves the efforts of both Cloud providers and their customers, whether they are individuals or individual companies, small, medium or large.

Cloud providers support their customers' services via network connections that may be carrier, network provider or the Internet. Because their business relies on customer trust, Cloud security methods must ensure that customer data remain private and are stored securely. However, Cloud security also relies in part on customers. Specifically, Cloud security relies on the following:

– data security;

– identity and access management;

– governance (policies for threat prevention, detection and mitigation);

– data retention and business continuity;

– legal compliance.

Cloud security may appear to be an extension of IT security, but in fact, a different approach must be considered which can be defined as the technology, protocols and practices that are capable of protecting Cloud Computing and Cloud Networking environments as well as the applications running in the Cloud and their data. Securing the Cloud requires a good understanding of what needs to be secured, as well as the system aspects that need to be controlled, managed and monitored.

The full scope of Cloud security is designed to protect the following:

– physical networks – routers, electrical power, cabling, climate controls, etc.;

– data storage – hard drives, etc.;

– data servers – basic network hardware and software;

– Computer Virtualization Frameworks – Virtual Machine Software, Host Machines and Guest Machines;

– Operating Systems (OS) – software that houses;

– data – all information stored, modified and accessed;

– applications – traditional software services (email, tax software, productivity suites, etc.);

– end user hardware – computers, mobile devices, Internet of Things (IoT), appliances, etc.

In Cloud Networking, the infrastructure owners can be very diverse and this needs to be taken into account. This can carry a lot of weight in terms of who is responsible for infrastructure and customer security.

We have seen in Chapter 6 that SASE (Secure Access Service Edge) was one of the solutions to secure the SD-WAN (Software-Defined Wide Area Network). SASE can be extended by taking into account Edge and Cloud parts and the customers associated with them. Depending on the use case, there may be many new elements to add to a SASE solution, and what is included in the solution may vary considerably depending on the vendor. But as a general rule, each SASI architecture provides elements of the five main components we already described in Chapter 6:

– SD-WAN: this technology, using SDN (Software-Defined Networking) to control WANs and their extension to the user, that is, the access network and the local network, is the foundation of SASE.

– CASB (Cloud Access Security Broker): according to Gartner, CASBS are on-premises or Cloud security policy enforcement points placed between Cloud service consumers and Cloud service providers to combine and intervene with enterprise security policies as Cloud resources are accessed.

– ZTNA (Zero Trust Network Access): ZTNA is a service that restricts identity and context-based access around an application or set of applications. Applications provide an environment that restricts access even to clients who appear to be legitimate.

– FWAAS (Firewall as a Service): FWAAS is a virtualized firewall in the Cloud to defend the network against cyber-attacks with the use of various filtering and threat prevention measures.

– SWG (Secure Web Gateway): SWGs filter and prevent data breaches on the network by applying new filtering measures on web access.

Edge and Cloud Networking can be secured in different ways depending on who is delivering the services. If Cloud services are implemented by vendors independent of the company, depending on the type of service the components to be managed are different.

First, we need to look at the provider of the physical network, data storage, data servers and virtualization architecture. We also need to look at which level of the architecture is supported by the vendor. The three classic levels (that we introduced in Chapter 2) are the following:

– Software as a Service (SaaS) provides customers with access to applications hosted and running on the provider's servers. Providers manage the applications, data, execution, middleware and operating system. Customers are only responsible for getting their applications.

– The PaaS (Platform as a Service) solution provides customers with a host to develop their own applications, which are run in the customer's sandbox services on the provider's servers. The provider must manage the runtime, middleware and operating system. Customers are responsible for managing their applications, data, user access and devices.

– The Infrastructure as a Service (IaaS) solution offers customers hardware environments all the way up to the operating system to support the company's computing, storage and network. Customers are responsible for securing everything above the operating system, including applications, data, middleware and the OS. In addition, clients must manage their user access, devices and the local network.

Cloud environments are deployment models in which one or more Cloud services create a system for end users and organizations. These services require management responsibilities, including security, which must be distributed between customers and providers. The types of Clouds currently in use include:

– Public Clouds composed of services shared between customers. These are third-party services managed by the provider, to provide access to customers via the Web.

– External private Clouds that are based on the use of a data center used by a single company. The data center is managed and operated by an external provider.

– Internal private Clouds composed of one or more data centers belonging to the company itself. In this case, the data center is controlled, managed and operated by the company. This allows a complete configuration of all the elements forming the company.

– Multi Clouds including the use of two or more Clouds from separate providers. These can bring any mix of public Clouds, external private Clouds and internal private Clouds.

– Hybrid Clouds being a mix of external private Clouds, internal private Clouds and one or more public Clouds.

Security in Cloud environments is different depending on the type of Cloud, especially if the Cloud is public or private. Each Cloud security measure works to accomplish one or more of the following:

– enabling data recovery in case of data loss;

– protecting storage and networks against data theft;

– preventing human error or negligence that causes data leakage;

– reducing the impact of any data or system compromise.

Data security is one aspect of Cloud security, which involves implementing techniques for threat prevention. Tools and technologies allow vendors and customers to insert barriers between access and visibility of sensitive data. Among these, encryption is one of the most powerful tools available. Encryption scrambles data so that it can only be read by someone who has the decryption key, which may be the same as the encryption key in the symmetric case. If the data are lost or stolen, they will be unreadable and meaningless. The protection during the transmission of data to access the Cloud can be done by a VPN (Virtual Private Network) technique.

Identity and Access Management (IAM) is about the accessibility privileges offered to user accounts. Authentication and authorization management of user accounts also applies here. Access controls are critical to restricting users – both legitimate and malicious – from entering and compromising sensitive data and systems. Password management, multi-factor authentication and other methods should be implemented to improve identity and access management.

Governance focuses on threat prevention, detection and mitigation policies. Threats should be tracked and prioritized to respond first to attacks on the most sensitive systems. However, even individual Cloud customers should undergo training on user behavior in the face of a failure. These forms of training are mostly applied in large organizations and less so in small ones, yet the rules to be put in place in the face of threats are almost the same regardless of the size of the company ranging from very large to a single individual user.

Protecting data and ensuring the resilience of ongoing operations involves technical disaster recovery measures. This requires the use of data redundancy methods such as backups. It also requires systems that ensure continuity of operations even in the event of failures or breakdowns. Software to test the validity of backups and recovery instructions are also very helpful in setting up a restart plan in case of a problem.

Legal compliance revolves around the protection of user privacy, as defined in many countries by legislative bodies. Governments have become aware of the importance of protecting private user information. As a result, companies have to follow all regulations to comply with these policies. One approach is the use of data masking techniques, which makes identity impossible to determine. This data masking is done through encryption methods.

Traditional IT security has evolved significantly due to the shift to Cloud Computing. While Cloud models make it easier to secure the environment, even though new attacks are emerging, connectivity requires support for new attacks that can extend into the data center. Cloud computing security, through the use of new technologies, is quite different from traditional IT legacy models on a number of points such as storage, scaling, interfaces and networking. We will now look at these four points first.

For data storage, the biggest distinction was the location of the storage, which was done on-premises instead of being externalized to the Cloud. Companies have long considered walls to be a barrier to accessing data storage with a good firewall. This is no longer true as mobile connections allow many people inside the company to connect to the outside with all the gateways available. The integration of data in the Cloud has made it possible to lower storage costs, increase security in certain areas, but at the same time remove a certain amount of user control.

Scaling, if it is well designed and only an extension of an established security framework, is not an issue. In contrast, Cloud security requires special attention when scaling IT systems. Cloud-centric infrastructures and applications are highly modular and quick to mobilize. This advantage allows the system to be adjusted to organizational changes. But this poses many problems when the upgrade requires the use of new paradigms. In this case, a new security framework must be adjusted or even designed.

For organizations and individual users, Cloud-based systems interfere with many other systems and services that must be secured. Access permissions must be developed to support all hardware and software from user devices to system access and with consideration of the network level. Beyond that, vendors and users need to be very mindful of the vulnerabilities they could cause by putting in an unsecured configuration or having abnormal system access behaviors.

Cloud-based systems involve a persistent link between Cloud providers and all users. The network can carry attacks going either way, to compromise customers or the providers themselves. In network environments, a single weak device or component can be exploited to infect every connected thing. Cloud providers expose themselves to threats from many end users as they interact with the Cloud. Network security responsibilities lie with the equipment suppliers but more often with the providers who released the product.

Solving most of the security issues posed by the Cloud requires that Cloud users and providers remain proactive about their cybersecurity roles. This two-pronged approach means that users and providers must agree and complement each other. To do this, Cloud providers and users need to be transparent with each other and share responsibilities fairly to keep the Cloud environment highly secure. These risks come from different horizons.

The first risk comes from the Cloud infrastructure, which must take into account risks from the IT environment that may be inherited from multiple generations on top of each other, and from storage service interruptions that may come from external data. The second risk is internal threats due to human error such as misconfiguration or poor user access controls. Next are external threats caused almost exclusively by malicious actors, such as phishing or DDOS. Finally, the biggest risk with the Cloud comes from an often ill-defined perimeter. Traditional cybersecurity has focused on protecting a well-defined perimeter, but Cloud environments are highly connected, which means that APIs (application programming interfaces) and attacks to successfully hijack an account can pose real problems. In the face of Cloud security risks, cybersecurity professionals must shift to a data-centric approach.

Interconnection also poses problems for the network side. Malicious actors often break into networks through the use of compromised or weak credentials. Once a hacker has managed to break into the network, they can easily develop and use poorly protected Cloud interfaces to locate the data they seek. They can even use their own Cloud servers as a destination where they can export and store all the stolen data. Security must be in the Cloud, not just in protecting access to data in the Cloud.

Third-party data storage and access via the Internet also contribute to the threats. If, for any reason, these services are interrupted, access to data is usually lost. For example, an Internet access outage could mean that access to the Cloud is impossible at a critical time. Alternatively, a power outage could affect the data center where the data are stored, possibly with permanent data loss. These interruptions could have long-term implications. A fire in a data center can result in data loss for some customers when the hardware is severely damaged. This is why local backups are essential for at least some critical data and for important applications. The reason for Edge computing in parallel with Cloud Computing is a good solution but requires even more security.

The introduction of Edge technology has forced a re-evaluation of cybersecurity. Data and applications migrate between local and remote systems and must always be accessible over the Internet or corporate network. As a result, protection is becoming more complex compared to Cloud computing alone where it was a matter of stopping unwanted users from accessing the Cloud data center.

Security has become essential for two key reasons. Cloud Computing allows the delocalization of a company's personnel who may be in the company, at home or traveling. The birth of SASE technologies came from there to protect all the accesses. Accesses can come from the company, from the data centers but also from the homes and from the place where the person who accesses the data and the applications are located. Of course, it is necessary to take into account the sharing of networks and data centers by multiple companies simultaneously in order to lower costs 24 hours a day.

In virtually every state, legislation has been put in place to protect users from having their sensitive data sold and shared. Examples in Europe include the GDPR (General Data Protection Regulation), portability laws and security requirements that Cloud providers must meet, but many states do not want to enter into such a legislation to allow more freedom for companies.

14.5. Edge Networking security

Edge Networking is presented on several levels: the MEC (Multi-access Edge Computing) level, the Fog level and the Embedded Edge level. In the rest of this section, we will look at the security of these different levels.

14.5.1. *Security of 5G MEC*

The main idea behind MEC is to move the computation of traffic and services from a centralized Cloud to the network edge and closer to the end user. Instead of sending all data to a Cloud for processing, network edge devices process and store data, reducing latency and bringing real-time performance to bandwidth-intensive applications. However, because MEC systems operate as open systems capable of running third-party application programs based on Cloud and virtualization technologies, they can be a major target for security threats, and the development of detection and response technologies is needed.

In the threat landscape, only three specific attacks against MEC have been mentioned: a fake or rogue gateway, overloading of edge nodes and abuse of open APIs, which can result in man-in-the-middle, flooding, and denial of service attacks, respectively. However, there are other vulnerabilities that open a new avenue of cyber threats as highlighted in many research papers. Among these works, Li et al. (2019a) showed that the 5G architecture allows attackers to launch a distributed DOS (DDOS) in a small area to prevent MEC service. In such an attack scenario, firewalls and intrusion protection systems (IPS) are essential for MEC nodes. However, their deployment can be a challenging task due to the economic and spatial limitations associated with the MEC environment.

According to 5G America, MEC systems operate as open systems capable of running third-party Cloud-based application programs. These programs introduce new threat vectors because they use open-source code, more interfaces and new APIs. In fact, many 5G providers and operators rely on valuable or reusable open-source software components to accelerate the delivery of digital innovation. As a result, some organizations may not have accurate inventories of software dependencies used by their various applications, nor a process to manage notifications about discovered vulnerabilities or available patches from the open-source supporting community.

Another category of security threats against MEC comes from resource misuse, where attackers leverage MEC resources to target users, organizations or other service providers. Threat actors may be a large-scale automated click fraud, stolen ID database brute force computer attacks, or perhaps digital currency mining.

14.5.2. Threats to Network Functions Virtualization

SDN and NFV (Network Functions Virtualization) are the key pillars of future networks, including 5G and much later 6G. NFV frees the network from its anchor in hardware and implements virtual networks on the common digital infrastructure. The main scope of this new concept is to make the system more dynamic and enable a centralized control plane, to easily integrate intelligence, without the need for human intervention or specific hardware configuration. NFV and SDN open the door to agile, flexible networks and the rapid creation of new services. While they incorporate security features, they also introduce additional challenges and complexities.

Many threats against virtualization technology have already been described, especially those related to data privacy. These threats can come from poor implementation of network configuration or exploitation of hypervisor vulnerabilities. There may also be misconfiguration of policy rules or poor isolation of containers or unprotected access to data belonging to other entities. Disclosure of sensitive information can also occur in virtual environments, where physical resources are shared between different companies. Another potential threat to NFV can come from abuse of resources and data centers. Virtual machines can be attacked by DOS or DDOS and may be susceptible to a power source attack. For example, if an attacker knows the scheduling characteristics of the hypervisor, the attacker can use this information to overload the resources, which performs a performance degradation of virtual machines (VMs).

In addition, the threat surface on NFV can greatly increase with the overlay of different threats on virtualization techniques (memory leakage, interrupted isolation, unscheduled interruption, etc.), combined with generic network attacks (flooding attacks, attack on routing or DNS, etc.).

Another attack related to the virtualization technique comes from the hypervision or containerization solution. In these attack scenarios, the attacker exploits periods in which the hypervisor or container suspends the VMs. This happens in hypervision when the hypervisor wants to get a consistent view of the hardware state. By determining how often the hypervisor pauses the virtual machine for inspection, the attacker can perform operations between monitoring controls. This allows the attacker, for example, to stealthily exfiltrate data or set up a backdoor on the virtual machine. In addition, because virtual computers are frequently instantiated (i.e. turned on and off), malware can propagate throughout the network by jumping from one virtual machine to another or from a VM on one host to many other hosts.

Other attacks can come from the use of free software. If the advantage is to be able to detect backdoors by examining the code, there are also many programming bugs that allow attacks. So, we should try to debug open-source software as much as possible.

14.5.3. *Fog security*

Fog security issues stem primarily from the many devices that are connected to Fog data centers and the intermediate gateways that are often required to transform the protocols used by objects into IP environments. Authentication plays a major role in establishing connections with IoT objects. This comes from these intermediate nodes not being secure enough to stop attacks at the device level which are quite regularly poorly protected. A lot of the big attacks in the Internet come from the IoT. For example, 450,000 video cameras, many of them baby monitors, were hacked by attackers who were able to send all the streams to a single point causing a flood that shut down a large part of the Internet in the United States and even slowed it down on a global scale.

In Fog computing, privacy preservation is even more complex because Fog data centers can collect sensitive data. In addition, because the nodes controlling Fog computing are scattered over large areas, a more distributed service must be implemented than the existing ones that are essentially centralized. Although the IoT is booming, this network offers a wide variety of services to users. It still faces many security and privacy issues.

IoT devices are connected to desktops or laptops in everyday life. The lack of security increases the risk of personal information leakage, while data are collected and routed either to the data center or to the object. In addition, IoT devices are often connected to different networks. Thus, if the IoT device has security holes, it can be the source of attacks that can be traced back to the data centers. The attack can then spread to other systems and damage them.

The main problem remains privacy and leakage of information harmful to users via their IoT devices. Not only can data be retrieved but also many other elements such as geographical location or actions performed.

With Fog computing, the data storage is local and unfortunately, being local does not protect the information unless the system has been put in containment so that nothing can get out or in. Fog computing improves latency and responds to demand in very short time frames, which can make it easier to detect attacks.

Security issues are significant in the Cloud and can be minimized by placing them in the Edge, through greater distance for attackers and better security through the use of VLANs, for example.

14.5.4. *Protection of intelligent processes in the Edge*

Attacks against embedded AI systems are on the rise due to the number of vulnerabilities discovered in these systems and the large-scale use of AI that we see everywhere, both for object defense and larger devices. These attacks can be classified into three broad categories. The first is conflicting data, which add noise and modify inputs so that they are misclassified or completely ignored during machine learning. Data modification is another attack that targets the training data. Samples are mislabeled so that the model learns in the wrong way and compromises the system for which it was set up. Another attack is carried out by modifying the learning model through a direct approach using complex engineering capable of modifying the model. Embedded artificial intelligence is often implemented in a wide range of use cases, from data acquisition to autonomous driving, which is why these rather sneaky attacks are really dangerous.

The list of vulnerabilities is long in embedded artificial intelligence production, as not only the data but also the algorithm, the model architecture of the system on which it is deployed is a potential target for attacks. To solve this problem, Cloud-based artificial intelligence services have been introduced, but they do not make security fool proof, because in some real-time applications, the network accessing the Cloud may not be reliable, making the system vulnerable. Therefore, artificial intelligence processes need to be heavily protected. They are now the target of many quite sophisticated attacks while waiting for solutions to counter them.

There is also a growing target of chips and boards. For example, an attacker can change a security parameter directly on the board that carries the artificial intelligence process, which will eventually cause the system to malfunction. These parameters control the security elements of the embedded AI system or peripherals, which is why they are so dangerous. In addition to physical attacks, an attacker can also create erroneous inputs that will result in the artificial intelligence system not learning properly.

Finally, some artificial intelligence applications at the Edge continue to learn from new data that continue to arrive in real time to improve the system and enhance responses. These applications are particularly vulnerable because they tend to update their detection parameters based on new learning data, making it much easier for the attacker to change access and gain access to the system.

One possible way to secure these systems is to equip them with a secure element such as a SIM card and, more and more often, an iSIM (integrated SIM) or an HSM, which we will examine in detail in the next section. These secure elements serve to keep sensitive information that resides directly in the SoC safe. They are used to protect data integrity and ensure that information is only accessible to authorized users or applications. In OTT (Over the Top) artificial intelligence systems (i.e. those added on top of the hardware and software environment of the device), the most important features to protect are models during the learning phase. These models must be protected by a secure environment, without leakage of sensitive information. A secure element will be able to guarantee that the parameters related to artificial intelligence are stored securely so that the calculation and detection are performed correctly and independently from other processes that use the same equipment.

14.5.5. *Client security through the use of HSM*

Client security is increasingly being introduced at Cloud providers through the addition of HSMs. HSMs are specialized hardware servers that give users the ability to obtain security tailored to the world of finance and cryptocurrencies. An HSM is a physical computing device that protects and manages digital keys, performs encryption and decryption of digital signatures, performs strong authentication and has many other cryptographic functions.

HSMs are critical to the security of the digital world and therefore to Cloud providers. They act as trusted systems that protect the encryption and cryptographic infrastructure of organizations that want very high security by supporting the secure management, processing and storage of encryption keys. Without them, data, transactions, and many jobs done in the Cloud could not be strongly protected.

This is why the largest Cloud service providers such as AWS, Google, IBM or Microsoft are now offering HSM services in their Clouds, to allow companies to keep data in their Clouds in a particularly protected way. AWS was the first to enter this market, followed by Google, then IBM and Microsoft. The extension of storage to data from high value-added services has been decisive in moving toward the same level of security as that used in finance and cryptocurrency processing.

Most companies running their IT jobs and applications in the Cloud were not using this same opportunity for their high-security applications. In fact, this view changed as of 2020, and by 2022 a high proportion of enterprises are relying on HSMs from Cloud providers.

14.6. Conclusion

The world of security is vast, and this chapter shows only a very partial view of network security. We have mainly focused on the new generation of security Clouds but also on new attacks, especially on AI processes. The Security Cloud has its advantages and disadvantages, but it is positioned on a new ground, that of simplification for companies that can manage all their security problems simultaneously. In addition, it simplifies access to security and is easily customizable to meet business objectives. The future will tell us how much of the market it will capture.

Another solution we discussed comes from blockchain, which is increasingly deployed and will certainly become a core technique for certification and traceability. Blockchain has the particularity of being a system without a central point. The transactions that carry the information to be traced; for example, a cryptocurrency purchase or a university degree certificate, and are collected in blocks. These blocks are encrypted against each other in such a way that if a transaction is touched, all blocks are modified and the discovery of the modification is almost immediate. This modification is then not taken into account so that all the blocks remain coherent. One of the difficulties of this distributed environment is choosing who encrypts the blocks because they could undo the block sequence and thus modify the transactions. For this, a consensus is needed to designate the leader who is in charge of the encryption of a block. There are many ways to elect this leader, for example, based on who has the most to lose by altering the blockchain. In Bitcoin, consensus is reached by a proof of work, which requires an extremely complex random calculation, and the winner is rewarded with a large amount of Bitcoin. This solution is criticized for being very expensive and for pushing the development of data centers that only perform one operation, the hash, since the objective is precisely to find the right value that gives a particular hash. In 2022, the world power to mine bitcoins was 200 Th/s (Tera-hashes per second), which consumes the electrical energy of a country like Belgium.

14.7. References

Andreoni Lopez, M., Mattos, D.M.F., Duarte, O.C.M.B., Pujolle, G. (2019). A fast-unsupervised preprocessing method for network monitoring. *Annals of Telecommunications*, 74(3–4), 139–155.

Belotti, M., Kirati, S., Secci, S. (2018). Bitcoin pool-hopping detection. *4th IEEE International Forum on Research and Technologies for Society and Industry IEEE RTSI*, Palermo, Italy.

Borisov, N., Goldberg, I., Wagner, D. (2002). Intercepting mobile communications. In *The Insecurity of 802.11, ACM Annual International Conference on Mobile Computing and Networking (MOBICOM)*, New York, USA.

Bozic, N., Pujolle, G., Secci, S. (2016). A tutorial on blockchain and applications to secure network control-planes. *3rd Smart Cloud Networks & Systems (SCNS)*, Dubai, UAE.

Bozic, N., Pujolle, G., Secci, S. (2017). Securing virtual machine orchestration with blockchains. In *1st Cyber Security in Networking Conference (CSNet)*, Rio de Janeiro, Brazil.

Bragadeesh, S. and Arumugam, U. (2019). *A Conceptual Framework for Security and Privacy in Edge Computing*. Springer International Publishing, Cham, Switzerland.

Caprolu, M., Di Pietro, R., Lombardi, F., Raponi, S. (2019). Edge computing perspectives: Architectures, technologies, and open security issues. In *IEEE International Conference on Edge Computing (EDGE)*, Milan, Italy.

Chilukuri, S., Bollapragada, S., Kommineni, S., Chakravarthy, K. (2017). Rain Cloud: Cloudlet selection for effective cyber foraging. In *Wireless Communications and Networking Conference (WCNC'17)*, IEEE, Raleigh, NC, USA.

Chwan-Hwa, J. and Irwin, J.D. (2017). *Introduction to Computer Networks and Cybersecurity*. CRC Press, Boca Raton, FL, USA.

Du, M., Wang, K., Xia, Z., Zhang, Y. (2020). Differential privacy preserving of training model in wireless big data with edge computing. *IEEE Transactions on Big Data*, 6(2), 283–295.

Fajjari, I., Aitsaadi, N., Dab, B., Pujolle, G. (2016). Novel adaptive virtual network embedding algorithm for Cloud's private backbone network. *Computer Communications*, 84(C), 12–24.

Gai, K., Wu, Y., Zhu, L., Xu, L., Zhang, Y. (2019). Permissioned blockchain and edge computing empowered privacy-preserving smart grid networks. *IEEE Internet of Things Journal*, 6(5), 7992–8004.

Goransson, P. and Black, C. (2016). *Software Defined Networks: A Comprehensive Approach*. Morgan Kaufmann, Burlington, MA, USA.

He, D., Chan, S., Guizani, M. (2018). Security in the internet of things supported by mobile edge computing. *IEEE Communications Magazine*, 56(8), 56–61.

Hu, Y., Perrig, A., Johnson, D. (2003). Packet leashes: A defense against wormhole attacks in wireless networks. In *IEEE Annual Conference on Computer Communications (INFOCOM)*.

Ibrahim, M.H. (2016). Octopus: An Edge-Fog mutual authentication scheme. *IJ Network Security*, 18(6), 1089–1101.

Kang, A., Yu, R., Huang, X., Wu, M., Maharjan, S., Xie, S., Zhang, Y. (2019). Blockchain for secure and efficient data sharing in vehicular edge computing and networks. *IEEE Internet of Things Journal*, 6(3), 4660–4670.

Khan, W.K., Aalsalem, M.Y., Khan, M.K., Arshad, Q. (2016). Enabling consumer trust upon acceptance of IoT technologies through security and privacy model. In *Advanced Multimedia and Ubiquitous Engineering*, Park, J.I., Jin, H., Jeong, Y.-S., Khan, M.M. (eds). Springer, Cham, Switzerland.

Khan, W.Z., Aalsalem, M.Y., Khan, M.K., Arshad, Q. (2019a). Data and privacy: Getting consumers to trust products enabled by the internet of things. *IEEE Consumer Electronics Magazine*, 8(2), 35–38.

Khan, W.Z., Ahmed, E., Hakak, S., Yaqoob, I., Ahmed, A. (2019b). Edge computing: A survey. *Future Generation Computer Systems*, 97, 219–235.

Khatoun, R. and Zeadally, S. (2017). Cybersecurity and privacy solutions in smart cities. *IEEE Communications Magazine*, 55(3), 51–59.

Lamport, L., Shostak, R., Pease, M. (1982). The byzantine general problem. *ACM Transactions on Programming Languages and Systems (TOPLAS)*, 4(3), 382–401.

Laubé, A., Martin, S., Al Agha, K. (2019). A solution to the split & merge problem for blockchain-based applications in ad hoc networks. In *International Conference on Performance Evaluation and Modeling in Wired and Wireless Networks*.

Li, Q., Meng, S., Zhang, S., Hou, J., Qi, L. (2019a). Complex attack linkage decision-making in edge computing networks. *IEEE Access*, 7, 12058–12072.

Li, X., Liu, S., Wu, F., Kumari, S., Rodrigues, J.J.P.C. (2019b). Privacy preserving data aggregation scheme for mobile edge computing assisted IoT applications. *IEEE Internet of Things Journal*, 6(3), 4755–4763.

Liu, M., Yu, F.R., Teng, Y., Leung, V.C., Song, M. (2019). Distributed resource allocation in blockchain-based video streaming systems with mobile edge computing. *IEEE Transactions on Wireless Communications*, 18(1), 695–708.

Ma, L., Liu, X., Pei, Q., Xiang, Y. (2019). Privacy-preserving reputation management for edge computing enhanced mobile crowdsensing. *IEEE Transactions on Services Computing*, 12(5), 786–799.

Mosenia, A. and Jha, N.K. (2017). A comprehensive study of security of internet-of-things. *IEEE Transactions on Emerging Topics in Computing*, 5(4), 586–602.

Nakamoto, S. (2008). Bitcoin: A peer-to-peer electronic cash system [Online]. Available at: http://www.bitcoin.org/bitcoin.pdf.

Nakkar, M., Al Tawy, R., Youssef, A. (2021). Lightweight broadcast authentication protocol for edge-based applications. *IEEE Internet of Things Journal*, 7(12), 11766–11777.

Ni, J., Lin, X., Shen, X.S. (2019). Toward edge-assisted internet of things: From security and efficiency perspectives. *IEEE Network*, 33(2), 50–57.

Nogueira, M., da Silva, H., Santos, A., Pujolle, G. (2012). A security management architecture for supporting routing services on WANETs. *IEEE Transactions on Network and Service Management*, 9(2), 156–168.

Onieva, J.A., Rios, R., Roman, R., Lopez, J. (2019). Edge-assisted vehicular networks security. *IEEE Internet of Things Journal*, 6(5), 8038–8045.

Rapuzzi, R. and Repetto, M. (2018). Building situational awareness for network threats in fog/edge computing: Emerging paradigms beyond the security perimeter model. *Future Generation Computer Systems*, 85, 235–249.

Rathore, S., Kwon, B.W., Park, J.H. (2019). Blockseciotnet: Blockchain-based decentralized security architecture for IoT network. *Journal of Network and Computer Applications*, 143, 167–177.

Roman, R., Lopez, J., Mambo, M. (2018). Mobile Edge computing, Fog et al.: A survey and analysis of security threats and challenges. *Future Generation Computer Systems*, 78, 680–698.

Salah, K., Rehman, M.H.U., Nizamuddin, N., Al-Fuqaha, A. (2019). Blockchain for AI: Review and open research challenges. *IEEE Access*, 7, 10127–10149.

Schneider, F. (1990). Implementing fault-tolerant services using the state machine approach: A tutorial. *ACM Comput.*, 22(4), 299–319.

Stiti, O., Braham, O., Pujolle, G. (2014). Creation of virtual Wi-Fi access point and secured Wi-Fi pairing, through NFC. *International Journal of Communication Networks, and Distributed Systems*, 12(7), 175–180.

Stojmenovic, J. and Wen, S. (2014). The fog computing paradigm: Scenarios and security issues. In *Computer Science and Information Systems (FedCSIS), Federated Conference*, IEEE.

Torres, J., Nogueira, M., Pujolle, G. (2013). A survey on identity management for the future network. *IEEE Communications Surveys and Tutorials*, 15(2), 787–802.

Wang, T., Zhang, G., Liu, A., Bhuiyan, M.Z.A., Jin, Q. (2019). A secure IoT service architecture with an efficient balance dynamics based on Cloud and Edge computing. *IEEE Internet of Things Journal*, 6(3), 4831–4843.

Wang, B., Li, M., Jin, X., Guo, C. (2020a). A reliable IoT edge computing trust management mechanism for smart cities. *IEEE Access*, 8, 373–399.

Wang, J., Wu, L., Choo, K., He, D.R. (2020b). Blockchain-based anonymous authentication with key management for smart grid edge computing infrastructure. *IEEE Transactions on Industrial Informatics*, 16(3), 1984–1992.

White, R. and Banks, E. (2018). *Computer Networking Problems and Solutions: An Innovative Approach to Building Resilient, Modern Networks*. Addison-Wesley, Boston, MA, USA.

Xiao, L., Ding, Y., Jiang, D., Huang, J., Wang, D., Li, J., Poor, H.V. (2020). A reinforcement learning and blockchain-based trust mechanism for edge networks. *IEEE Transactions on Communications*, 68(9), 5460–5470.

Yang, Y., Wu, L., Yin, G., Li, L., Zhao, H. (2017). A survey on security and privacy issues in Internet-of-Things. *IEEE Internet of Things Journal*, 4(5), 1250–1258.

Yang, J., Lu, Z., Wu, J. (2018). Smart-toy-Edge-computing-oriented data exchange based on blockchain. *Journal of Systems Architecture*, 87, 36–48.

Yang, R., Yu, F.R., Si, P., Yang, Z., Zhang, Y. (2019). Integrated blockchain and Edge computing systems: A survey, some research issues and challenges. *IEEE Communications Surveys Tutorials*, 21(2), 1508–1532.

Yeluri, R. and Castro-Leon, E. (2014). *Building the Infrastructure for Cloud Security: A Solutions View*. Apress Open (Springer Nature), Cham, Switzerland.

Yi, S., Qin, Z., Li, Q. (2015). Security and privacy issues of fog computing: A survey. *International Conference on Wireless Algorithms, Systems and Applications (WASA)*.

Yuan, J. and Li, X. (2018). A reliable and lightweight trust computing mechanism for IoT edge devices based on multi-source feedback information fusion. *IEEE Access*, 6, 23626–23638.

Zhang, H., Hao, J., Li, X. (2020). A method for deploying distributed denial of service attack defense strategies on edge servers using reinforcement learning. *IEEE Access*, 8, 482–491.

15

Accelerators

A first solution to accelerate software processing comes from accelerators. Many performance accelerator products are already on the market such as DPDKs (Data Plane Development Kits). The DPDK environment is a set of data plane libraries and network interface drivers that accelerate communications. Actual virtual machine migrations – the actual transport of all the software associated with a virtual machine – from one physical machine to another physical machine now take only a few milliseconds to complete.

A second accelerator, FD.io (Fast Data – Input/Output), comes from a simple observation that the more congestion there is, the more packets are waiting in the output queues and the more time can be saved by processing several packets from the same session simultaneously.

DPDK and FD.io are good examples of accelerators that can bridge the performance gap of processors handling certain software that needs to achieve high performance. There are other types of accelerators that can be seen as intermediaries between pure software solutions and pure hardware solutions. However, the advent of increasingly powerful reconfigurable microprocessors is a much better way to accelerate processing.

The return to hardware is inevitable. Indeed, going back to hardware accelerates performances and gains in consumed energy, but the loss comes from a lower flexibility. To make up for this deficit, we need to move to hardware virtualization instead of software virtualization. We will examine these different acceleration solutions in different sections of this chapter.

15.1. The DPDK accelerator

The objective of DPDK is to create a set of libraries for environments that group together software that must perform specific functions requiring significant computing power. This type of software can be found in the context of virtualization for signal processing or multimedia protocol management applications. This solution is applied to the network domain through an additional layer called EAL (Environment Abstraction Layer). This EAL hides the specifics of the environment by providing a programming interface to libraries, interfaces, hardware accelerators and elements of operating systems such as Linux or FreeBSD. Once EAL is created for a specific environment, developers chain together libraries and interfaces to create their applications. For example, EAL can provide a standard framework to support Linux, FreeBSD, 32- or 64-bit Intel IA, or IBM Power8.

EAL also provides additional services such as timing references, PCIe (Peripheral Component Interconnect Express) bus access, monitoring functions, debugging solutions and alarm functions. Finally, the DPDK environment implements a model with very little overhead, which results in excellent performance on the data plane. The environment also allows access to individual nodes by eliminating processing overhead, which leads to quite surprising speedups that can be achieved with fairly simple processes. The DPDK environment also incorporates examples that highlight best practices for software architectures, tips for designing data structures and storage, application tuning, and tips for closing performance gaps in different network deployment solutions.

The communications industry is moving toward network architectures, operations and services based on standard servers and Cloud Native principles. This requires high levels of function partitioning and hardware abstraction to increase modularity and portability. Even with the collective progress made on NFV (Network Functions Virtualization), we still have work to do to implement very high-performance Cloud Native.

Some communication service providers face challenges when integrating and orchestrating virtual network functions (VNFs) in part because of hardware and software interdependencies. Data planes depend on hardware accelerators and specific platform configurations that are not Cloud Native. Evolving standards for I/O virtualization, such as VirtIO, have performance limitations. SRIOV (Single Root I/O Virtualization) addresses these performance requirements, but this comes at the cost of reduced flexibility, such as limiting migrations.

Complete disaggregation has not been possible for networks in which the layers are not completely independent of each other. As with many technological developments, there is no single approach.

Moving toward higher performance solutions requires removing data plane dependencies and ensuring that current hardware accelerators and new next-generation instruction optimizations are usable by VNFs. For example, Intel's approach is to integrate new packet processing capabilities into the data plane through virtualized application acceleration. This is to complement the acceleration potential of the infrastructure with appropriate network cards and SmartNICs. Intel believes that this platform partitioning enables infrastructure-related accelerations through network cards and SmartNICs. Application acceleration will provide the scalability required for the next generation of applications that will be increasingly data dependent.

Cloud Native applications require network services to be abstracted and easy to use without any platform dependencies, while providing higher throughput and lower latencies. The abstraction between hardware and software continues where data planes further evolve to incorporate industry standard network hardware, common APIs, and are supported by additional automation to handle decoupling of layered systems.

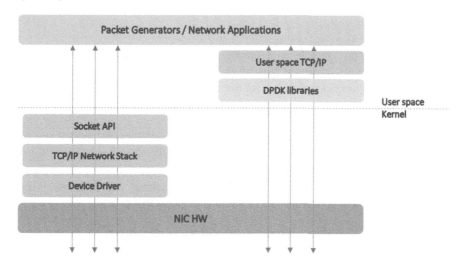

Figure 15.1. *Architecture using DPDK. For a color version of this figure, see www.iste.co.uk/haddadou/edge.zip*

The open-source community supports the evolution of the abstraction with the DPDK. Data plane VNFs require high-performance network interfaces, high online data rates and low latency, all at minimal cost. The DPDK is part of this evolution. The network card and hardware upgrades are intended to enable support for Cloud Native runtime environments. Figure 15.1 details the software architecture and DPDK involvement.

A virtual switch (vSwitch) enables communication between virtual machines (VMs) and containers. A classic option is offered by the Open vSwitch (OvS) which is an open-source implementation of a virtual switch and a critical component for network virtualization in Cloud computing. While Open vSwitch is a critical component of the Cloud, concerns exist about Cloud Networking performance and scalability requirements. By leveraging the libraries provided by the DPDK, OvS network performance weaknesses are addressed and CPU core utilization is optimized. As a result, infrastructure cost is minimized, and CPU core availability is maximized for Cloud service delivery.

Figure 15.2 describes the integration of DPDK into the OvS (Open vSwitch) architecture.

When a virtual switch incorporates DPDK, resulting in OvS-DPDK, performance is greatly improved over an original OvS machine. The metrics show this significant performance change over the standard scheme:

– network performance: throughput, latency and jitter;

– network stability: reliability and scalability of DPDK-based applications by running on the DPDK libraries. As a result, OvS-DPDK includes OvS Native features while significantly improving packet processing performance. In addition, DPDK-based Vhost user ports allow virtual machine compatibility. This compatibility enables support for live migrations to OvS-DPDK.

Compared to native OvS, the forwarding plane in OvS-DPDK is shifted from kernel to user space, which helps to improve system stability. OvS-DPDK uses PMDs (poll mode drivers) instead of kernel-installed, interrupt-based drivers as used in older OvS switches. By using a PMD for network packets, this eliminates interruptions to network traffic and removes service routines for capturing network data.

Eliminating interruptions helps improve network performance by avoiding particularly expensive context switches. In contrast, a PMD running in user space communicates directly with the hardware, bypassing the kernel completely. This mechanism not only reduces network latency by eliminating the overhead of copying data from the kernel to user space but also allows for better resource utilization by not allocating an intermediate buffer in the kernel. When implementing OvS-DPDKs, Cloud service providers are able to achieve significant performance advantages over traditional switches.

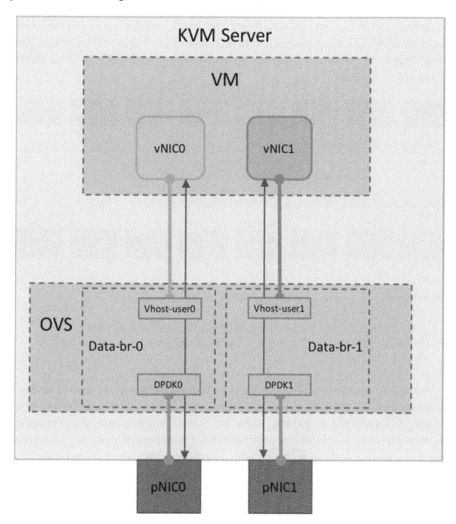

Figure 15.2. *The introduction of DPDK in OvS. For a color version of this figure, see www.iste.co.uk/haddadou/edge.zip*

15.2. The FD.io accelerator

The FD.io accelerator is based on a simple idea: the more congestion there is, the more packets are waiting in the nodes' input queues to be processed. As a result, the probability that there are multiple packets belonging to the same session increases. Since the time to process one or more packets in the node is almost the same, it is interesting to gather all packets from the same session and process them simultaneously. The more congestion there is, the more efficient the accelerator is. When there is no congestion, there is no acceleration possible. Figure 15.3 shows accelerator architecture.

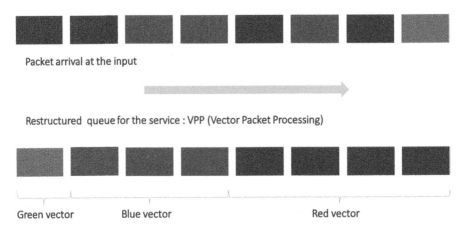

Figure 15.3. *The architecture of the FD.io accelerator. For a color version of this figure, see www.iste.co.uk/haddadou/edge.zip*

Figure 15.4 shows the gains made by the FD.io accelerator in the case of an SDN (Software-Defined Networking) using OpenFlow signaling. With a single packet to process, it is necessary to go back to the flow table, which is particularly complex and requires a very expensive processing time. The following packets of the same flow that are processed can go back to the megaflow cache, which takes a little time but is very marginal compared to the processing in the flow table. Finally, the following packets can use the microflow cache and in this case its processing is almost instantaneous. We can thus see that the time to process identical packets is approximately the processing time in the flow table.

Accelerators 259

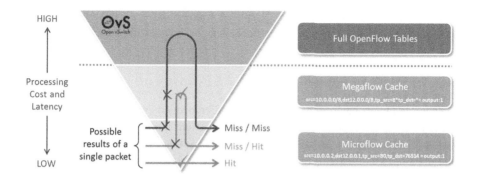

Figure 15.4. *FD.io intervention levels. For a color version of this figure, see www.iste.co.uk/haddadou/edge.zip*

FD.io relies heavily on a collection of tools primarily from DPDK. By greatly reducing CPU interrupt overhead, DPDK's PMDs periodically poll the pending packets and prepare for processing on vectors of packets belonging to the same session.

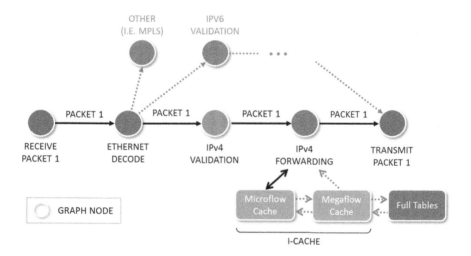

Figure 15.5. *Description of the FD.io solution. For a color version of this figure, see www.iste.co.uk/haddadou/edge.zip*

Figure 15.5 shows the transfer graph of frames and packets until they pass through either the flow table, the megaflow cache or the microflow cache.

15.3. Hardware virtualization

More and more heterogeneous ARM-based embedded systems are composed of powerful processors combined with energy-efficient programmable accelerators. Processor virtualization is a technology that is more mature than accelerator virtualization, this is due to the late introduction of extensions for virtualization in hardware accelerators. Virtualization decreases the overall cost by sharing hardware, and thus improves computational load balancing and system flexibility.

The latest improvements in accelerators that support hardware virtualization, combined with virtualization extensions for processors, allow for negligible performance loss compared to native performance. There are many types of accelerators that can be virtualized (such as GPUs, FPGAs [Field-Programmable Gate Array]). A common virtualization software architecture would therefore reduce software fragmentation but also reduce maintenance costs. Virtual Open Systems has developed a generic virtualization interface, based on VFIO, to gather all possible accelerators around the same virtualization architecture.

Figure 15.6 describes a common architecture for accelerator virtualization. The generic virtualization interface separates the virtualization part, which is open source, from the manufacturer part, which is accelerator specific (hardware driver). This architecture and its components are shown in Figure 15.6.

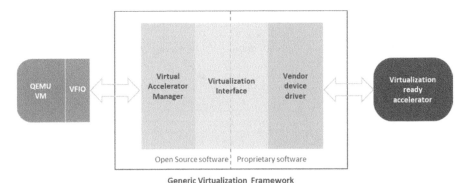

Figure 15.6. *Acceleration architecture. For a color version of this figure, see www.iste.co.uk/haddadou/edge.zip*

A reconfigurable processor is a microprocessor with hardware that can dynamically rewire itself. This allows the chip to efficiently adapt to the programming tasks required by a particular software program. For example, the reconfigurable processor can transform from a video chip to a graphics chip, both optimized to allow applications to run at the highest possible speed. We could say they are processors on demand. In a practical sense, this capability can translate into a great deal of flexibility in terms of the chip's functions. For example, a single chip could serve as a camera, signal processor, router and firewall. To do this, all that is required is to download the desired software, and the processor automatically reconfigures itself to optimize the performance required for the programmed function.

Several types of reconfigurable processors are available on the market: first, DSPs (Digital Signal Processor), which have excellent performance. These are programmable chips used in cell phones, automobiles, and various types of music and video players. Another version of reconfigurable microprocessors is equipped with programmable memory arrays that perform hardware functions using software tools. These microprocessors are more flexible than specialized DSP chips, but also slower and more expensive. Hardwired chips are the oldest, cheapest and fastest but are unfortunately the least flexible.

Reconfigurable microprocessors have different names depending on what is reconfigured. The best known are the FPGAs, which are arrays of programmable gates in RAM technology.

Reconfigurable microprocessors are a new generation of systems that should eventually replace all software that need some processing power. These reconfigurable microprocessors consist of a layer of hardware elements and a layer of memory elements. The hardware element is fairly complex, which represents granularity. The granularity is expressed from a fine to a thick version to express the complexity of the hardware element. For a fine granularity, we work at the bit level which enables programming any logic function. However, this solution is expensive in terms of performance and does not gain all the orders of magnitude that one would like for some processing components in networks. Moreover, the reconfiguration time is too long to handle several real-time processes on the same reconfigurable processor.

In the case of thick grains, not all functions can be realized directly. The elements form operators that can be used directly for operations required in signal processing or multimedia protocols. These operators can be reconfigured much faster because they are limited in the number of functions that can be taken into account by the overall component.

More precisely, the granularity of the reconfigurable element is defined as the size of the smallest basic block, the CLB (Configurable Logic Block), that can be put in the chain of functions to be realized. A high granularity, which we also introduced as a fine granularity, implies a great flexibility to implement the algorithms in hardware. Almost any type of function can be implemented. We use fine granularity to realize particular functions or to test new algorithms, before eventually moving on to thicker granularity. The difficulties with fine-grained circuits are the higher power to be supported and the lower execution speed due to the long path to be covered. Reconfiguration can also be time-consuming compared to the time required to maintain a real-time process. On the contrary, coarse grains have much shorter chaining paths and make it easier to approach real time.

Of course, the processing of a function must match the path as closely as possible. If the granularity is too thick, the component may take longer than necessary and therefore be poorly used, coupled with higher power consumption. For example, a 4-bit addition performed on a 16-bit granularity component loses performance while consuming significantly more power. The idea to find the best compromise is to realize arrays of thick-grained elements mixed with fine-grained elements. Such a scheme can be obtained on a single chip with rDPA (reconfigurable Datapath Array) and FPGA.

The architectures of rDPA and FPGA type arrays must be optimized so that the path between the different elements is as short as possible. If the architecture is not adapted, a strong loss of efficiency is immediately perceptible but is not necessarily a major defect if the global circuit is used for many algorithms that are very different from each other. In fact, the best thing is to know in advance the algorithms that will be executed on the circuit to optimize the logic elements, the memory and the path to follow. FPGAs are generally too thin to be very efficient in performance, and it is necessary to replace them with thicker elements adapted to the objectives of a reconfigurable microprocessor.

Reconfiguration can be performed either at circuit startup, between two execution phases or even during the execution of a process. Circuits working at the binary element level, such as FPGAs, require a significant amount of time for their reconfiguration, which is performed by a rather complex bitstream compared to thick-grained architectures that require a much shorter bitstream with a much faster implementation. Most reprogrammable boards actually have partial reconfigurations allowing the execution of a new function, while another part of the microprocessor continues to execute a previous function.

Reconfigurable element arrays seem to be the best compromise between fine and thick grain by putting together enough circuits to find optimal chaining paths. To control this array, a host processor is needed. It must be powerful to determine which algorithms to use to determine the chaining. Overall, the flexibility of the architecture comes from the path realized to interconnect the gates. The path to be determined is therefore of primary importance for the realization of complex functions such as those found in some protocol processing or signal processing.

15.4. Conclusion

With software virtualization, the loss of performance is very noticeable compared to hardware. This is due to the fact that everything that accelerates processing, such as ASICs (Application-Specific Integrated Circuits), has disappeared. To regain part of the performance loss, an optimized processing on the memory with the processor consists of setting up fast paths allowing accelerations up to a factor of 10 and at extremely low prices since most accelerators are open source. The two most used have been described previously (DPDK and FD.io). We have also shown that we will switch back to hardware virtualization by around 2028, but the flexibility will not be as important as in software. However, the gains in performance and power consumption will be enormous.

15.5. References

Baldini, I., Castro, P., Chang, K., Cheng, P., Fink, S., Ishakian, V., Mitchell, N., Muthusamy, V., Rabbah, R., Slominski, A. et al. (2017). Serverless computing: Current trends and open problems. In *Research Advances in Cloud Computing*, Chaudhary, S., Somani, G., Buyya, R. (eds). Springer, Singapore.

Belay, A., Prekas, G., Primorac, M., Klimovic, A., Grossman, S., Kozyrakis, C., Bugnion, E. (2016). The IX operating system: Combining low latency, high throughput, and efficiency in a protected dataplane. *ACM Transactions on Computer Systems* (TOCS), 34(4), 1–39.

Blenk, A., Basta, A., Reisslein, M., Kellerer, W. (2016). *Survey on Network Virtualization Hypervisors for Software Defined Networking*. IEEE Communications Surveys & Tutorials, 18(1), 655–685.

Bosshart, P., Gibb, G., Kim, H.-S., Varghese, G., McKeown, N., Izzard, M., Mujica, F., Horowitz, M. (2013). Forwarding metamorphosis: Fast programmable match-action processing in hardware for SDN. *ACM SIGCOMM*, 44(3), 87–95.

Caulfield, A., Costa, P., Ghobadi, M. (2018). Beyond SmartNICs: Towards a fully programmable Cloud. *IEEE 19th International Conference on High Performance Switching and Routing (HPSR)*.

Deri, L., Martinelli, M., Bujlow, T., Cardigliano, A. (2014). nDPI: Opensource high-speed deep packet inspection. *International Wireless Communications and Mobile Computing Conference*, Nicosia, Cyprus.

ETSI GS NFV-IFA 001 (2015). Report on acceleration technologies & use cases. ETSI Group Specification, Rev. 1.1.1.

Firestone, D. (2017). VFP: A virtual switch platform for host SDN in the public Cloud. *USENIX Symposium on Networked Systems Design and Implementation (NSDI '17)*, Boston, MA, USA.

Firestone, D., Putnam, A., Mundkur, S., Chiou, D., Dabagh, A., Andrewartha, M., Angepat, H., Bhanu, V., Caulfield, A., Chung, E. et al. (2018). Azure accelerated networking: SmartNICs in the public Cloud. *USENIX Symposium on Networked Systems Design and Implementation (NSDI)*, (NSDI'18), Renton, WA, USA.

Guan, H., Dong, Y., Ma, R., Xu, D., Zhang, Y., Li, J. (2013). Performance enhancement for network I/O virtualization with efficient interrupt coalescing and virtual receive-side scaling. *IEEE Transactions on Parallel and Distributed Systems*.

Han, S., Jang, K., Panda, A., Palkar, S., Han, D., Ratnasamy, S. (2015). SoftNIC: A software NIC to augment hardware. Technical Report, EECS Department, University of California, Berkeley, CA, USA.

Hoffmann, M., Jarschel, M., Pries, R., Schneider, P., Jukan, A., Bziuk, W., Gebert, S., Zinner, T., Tran-Gia, P. (2017). SDN and NFV as enabler for the distributed network cloud. *Mobile Networks and Applications*, 23(3), 521–528.

Hwang, J., Ramakrishnan, K.K., Wood, T. (2015). NetVM: High performance and flexible networking using virtualization on commodity platforms. *IEEE Transactions on Network and Service Management*, 12(1), 34–47.

Kourtis, M.A., Xilouris, G., Riccobene, V., McGrath, M.J., Petralia, G., Koumaras, H., Gardikis, G., Liberal, F. (2015). Enhancing VNF performance by exploiting SR-IOV and DPDK packet processing acceleration. *IEEE Conference on Network Function Virtualization and Software Defined Network*.

Lettier, G., Maffione, V., Rizzo, L. (2017). Survey of fast packet I/O technologies for network function virtualization. *International Conference on High Performance Computing*.

Li, S., Lim, H., Lee, V.W., Ahn, J.H., Kalia, A., Kaminsky, M., Andersen, D.G., Seongil, O., Lee, S., Dubey, P. (2015). Architecting to achieve a billion requests per second throughput on a single key-value store server platform. *Annual International Symposium on Computer Architecture*.

Li, X., Sethi, R., Kaminsky, M., Andersen, D.G., Freedman, M.J. (2016). Be fast, cheap and in control with switchKV. *USENIX Symposium on Networked Systems Design and Implementation (NSDI'16)*, Santa Clara, CA, USA.

Madhavapeddy, A. and Scott, D.J. (2014). Unikernels: The rise of the virtual library operating system. *Communications of the ACM*.

Marinos, I., Watson, R.N.M., Handley, M. (2014). Network stack specialization for performance. *ACM SIGCOMM*, 11(11), 177–185.

McGrath, G. and Brenner, P.R. (2017). Serverless computing: Design, implementation, and performance. *International Conference on Distributed Computing Systems Workshops*.

Pitaev, N., Falkner, M., Leivadeas, A.A., Lambadaris, I. (2018). Characterizing the performance of concurrent virtualized network functions with OVS-DPDK, FD.IO VPP and SR-IOV. *ACM/SPEC International Conference on Performance Engineering*.

Pontarelli, S., Bianchi, G., Welzl, M. (2018). A programmable hardware calendar for high resolution pacing. *High Performance Switching and Routing, IEEE 19th International Conference on SDN*.

Trevisan, M., Finamore, A., Mellia, M., Munafo, M., Rossi, D. (2017). Traffic analysis with off-the-shelf hardware: Challenges and lessons learned. *IEEE Communication Magazine*, 55(3), 163–169.

Wang, H., Soule, R., Dang, H.T., Lee, K.S., Shrivastav, V., Foster, N., Weatherspoon, H. (2017). P4FPGA: A rapid prototyping framework for P4. *Symposium on SDN Research, ACM SIGCOMM*, 122–134.

Yi, B., Wang, X., Li, K., Das, S., Huang, M., (2018). A comprehensive survey of network function virtualization. *Computer Networks*, 133, 212–262.

Zhang, Q., Liu, V., Zeng, H., Krishnamurthy, A. (2017). High-resolution measurement of data center microbursts. In *Proceedings of the Internet Measurement Conference*, ACM, London, UK.

Zilberman, N., Watts, P.M., Rotsos, C., Moore, A.W. (2015). Reconfigurable network systems and software-defined networking. *Proceedings of the IEEE*, 103(7).

16

The Future of Edge and Cloud Networking

The future of Edge and Cloud Networking can be imagined through the research done for 6G. This new generation seems far away, in fact its study started in 2020 and it will take almost a decade to arrive at a specification. The years 2020–2024 will be mainly devoted to the different possible ways to achieve a new generation.

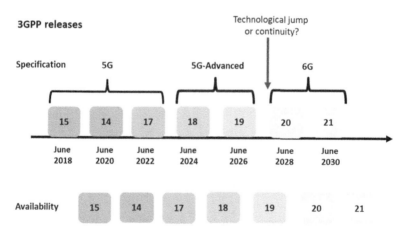

Figure 16.1. *Specifications for moving from 5G to 6G. For a color version of this figure, see www.iste.co.uk/haddadou/edge.zip*

Figure 16.1 shows the dates of the planned specifications, but they are obviously not definitive, far from it. On average, a new specification appears every 2 years. On this diagram, the first three releases 15, 16 and 17 correspond to the three phases of 5G. The first one concerns 5G radio, called NR (New Radio), that is, the part between the terminal equipment and the antenna. The second specifies the radio access network and the core network. Finally, the third phase is largely devoted to 5G applications.

The 5G revolution is represented by the appearance of data centers on the Edge: MEC (Multi-access Edge Computing) data centers. The objective of this architecture is to offer services positioned in the data center, which must be less than 10 km from the user to ensure a latency time of no more than 1 ms. This solution enables a whole new set of services to be offered from virtual machines, services that cannot be supported by current architectures that position data centers hundreds or even thousands of kilometers from the user.

Another important piece of information is the name of these data centers: MEC. Initially its meaning was Mobile Edge Computing, indicating that it was dedicated to connecting mobiles via an antenna. The new name given at the end of 2020 indicates that the data center can be accessed by different networks including networks that do not use antennas or an antenna at the terminal level like Wi-Fi.

The fifth generation is essentially characterized by its applications that run in a data center that is on the Edge.

Another major feature that will develop strongly in 6G is private networks and thus in this case private 5G. Businesses will increasingly have a local infrastructure based on 5G and Wi-Fi with the two technologies coming together through their control plane that can easily interconnect and allow coordination between Wi-Fi and 5G. This is expected to happen as soon as Wi-Fi 7 spreads, which is the first Wi-Fi network to have this feature.

In the description of 5G as it will develop, an important element that has a role in 6G is slices, which are virtual networks that are created to support particular applications or enterprise networks. This slice-based 5G network architecture is shown in Figure 16.2.

5G-Advanced offers strong virtualization as all functions will be virtualized in the MEC data center.

To move toward the future and 6G, we must describe the three main possible paths that are defended by the major telecom manufacturers and operators but also the major web and data center companies since, as we have just seen, the revolution in these systems comes from the appearance of data centers and their increasing use.

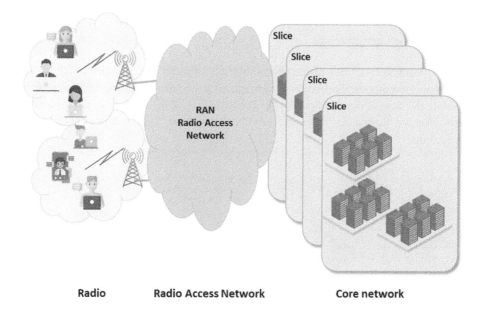

Figure 16.2. *A 5G network with slices. For a color version of this figure, see www.iste.co.uk/haddadou/edge.zip*

The three possible paths are as follows: a continuation of 5G by offering more antennas and more slices, a revolution by moving toward a fully distributed network where the data centers are embedded in the end devices, and finally, an intermediate solution where virtual machines can be positioned on a whole hierarchy of data centers. We will describe these three paths and show the evolution of data centers.

16.1. 5G continuity

The first solution is to continue down the path opened up by 5G. We keep the MEC data centers by amplifying the size of the data centers and pushing the Edge's border to 50 km from the user. The size of the data centers becomes huge, as this center can handle millions of users by connecting antennas via fiber to 50 km away. One of the goals of these data centers is to become highly intelligent with specialized processors, for example, toward neural techniques. A server can support up to a hundred neurons, and by grouping a billion servers, we reach the value of a hundred billion neurons which is the number in the human brain. To feed these billion servers will take 100 GW, equivalent all the energy produced instantaneously in a country like France. The brain consumes 20 W to do the same thing, maybe even better. In 2028, the energy consumption of a computer brain will be equal to that of 5 billion human brains.

The continuity of 5G also applies to slices. Indeed, the number of slices is increasing steadily in 5G, first by the horizontal slices corresponding to 5G services and all the new services, then by the vertical slices that will be set up to create networks for businesses. Figure 16.3 illustrates these slices, whose number is increasing.

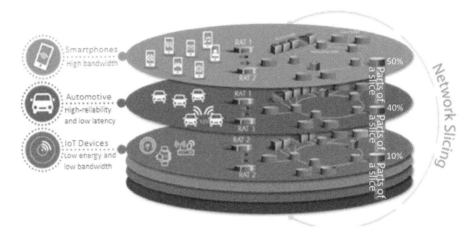

Figure 16.3. *Increasing the number of slices. For a color version of this figure, see www.iste.co.uk/haddadou/edge.zip*

6G will extend the number of slices to go toward a slice per user who wants to connect: at the time of the request of opening of the session, the corresponding virtual network is set up thanks to the opening of virtual machines carrying out the nodes of the network. At the end of the session, the virtual machines are turned off to destroy the virtual network. In fact, a user can have several virtual networks depending on the number of devices connected. The advantage of this solution is to open virtual networks completely customized to the application in terms of performance, quality of service and security. The disadvantage comes from the scaling up. The digital infrastructure must be able to support billions of virtual networks that may all be different from each other. The management of this set of networks will be extremely complex. However, each company will be able to design its own network, with its own virtual machines with the advantage of being more difficult to attack since the protocols of these networks will not be known.

Finally, to complete the picture of this new generation using the continuity of 5G, we must also mention the increase in the number of antennas. These antennas are very strongly directional. For 5G, this number will increase very soon: from 1,000 sub antennas, we could go to 100,000 subantennas, meaning we could put up to a million antennas on a single mast. So, the slice will contain both a part of the RAN (Radio Access Network) and a part of the core network. In conclusion, there will be a complete slice including the radio, the RAN and the core network per connection. In addition, these antennas will be hyper directional so that there will be only one connection on the antenna as shown in Figure 16.4.

Figure 16.4. *Connections via directional antennas. For a color version of this figure, see www.iste.co.uk/haddadou/edge.zip*

Cellular networks are disappearing to make way for point-to-point connections since the cell no longer exists.

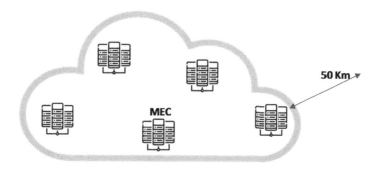

Figure 16.5. *6G with a vision of centralization. For a color version of this figure, see www.iste.co.uk/haddadou/edge.zip*

With a fairly high degree of centralization, the data centers become larger to connect all the antennas within 50 km. Indeed, the round-trip time gets a bit longer but remains in the microsecond range. There are 80 additional kilometers of fiber to travel, which at 200,000 km/s, which is the propagation speed of light in the fiber, represents a time of 400 μs. Figure 16.5 shows the 6G data center environment in a case where centralization is required.

16.2. Fully distributed networks

A second vision for 6G is to distribute MEC data centers as much as possible. Figure 16.6 symbolizes this research direction for 6G.

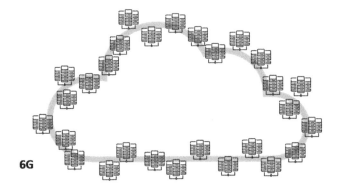

Figure 16.6. *Distribution of data centers at the Edge. For a color version of this figure, see www.iste.co.uk/haddadou/edge.zip*

For this, two solutions stand out: putting the data center in the end device (i.e. the object, tablet, smartphone, etc.) or putting an intermediate box (e.g. a 5G router) to multiplex the virtual machines of the Edge, that is, the application service virtual machines, infrastructure service virtual machines or digital infrastructure virtual machines.

In both cases, the data center is embedded in the terminal or in the intermediate box. If the data center is in the terminal machine, the connections are made in direct mode, that is, directly from one piece of terminal equipment to another, which is still called D2D (Device to Device). When the connection is at the user's machine or if it is an object, we are in an ad hoc mode. If the data center is located in a box not far from the end device, the solution is a mesh network with the boxes connected in direct mode. In this case, the boxes are the property of an infrastructure operator who must manage the security, control and management of this network.

D2D is extended in ad hoc or mesh mode if the nodes in the network have a flow table, which is usually a routing table. In this case, the communication works with hops from node to node, which is called multi-hop communication.

The first case is shown in Figure 16.7. The smartphones that contain the MEC data centers interconnect directly with each other via direct mode, but they can also connect via a satellite or a connection through a 4G or 5G infrastructure.

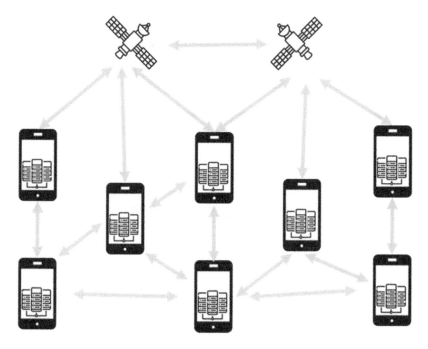

Figure 16.7. *A possible 6G ad hoc mode. For a color version of this figure, see www.iste.co.uk/haddadou/edge.zip*

This first distributed environment solution may not be the first to be deployed because of the difficulty of embedding a data center in a smartphone, tablet or even an object. The second possibility seems more realistic with a multiplexing of the virtual machines of the end devices in a box that has several antennas as shown in Figure 16.8. The main antenna allows the boxes to interconnect in mesh mode. The other antennas allow the end devices to connect to the box. These antennas are Wi-Fi and 5G for the main ones and Bluetooth, Wi-Fi HaLow, ZigBee, etc. for the others. 5G is in horizontal mode whereas the connections between boxes are in direct mode using 6G D2D, which will be an extension of 5G D2D.

Figure 16.8. *The box of a 6G mesh network (Green Communications). For a color version of this figure, see www.iste.co.uk/haddadou/edge.zip*

The boxes are interconnected through their 5G antenna in direct mode. They can also be interconnected through Wi-Fi and both at the same time, resulting in a hybrid network where the best route can alternatively go through 5G and Wi-Fi.

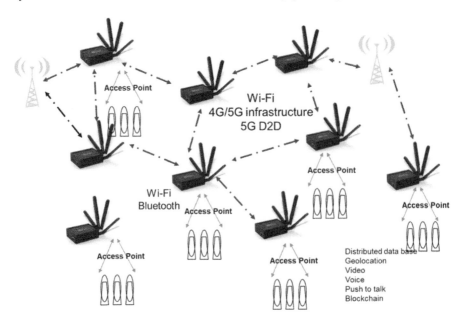

Figure 16.9. *A 6G mesh network (Green Communications). For a color version of this figure, see www.iste.co.uk/haddadou/edge.zip*

Figure 16.9 shows a 6G mesh network with boxes carrying an MEC data center supporting all virtual machines related to end machine applications, service applications such as security, management, automation, artificial intelligence and finally digital infrastructure. These boxes can also be called femto data centers.

Femto data centers can also be connected to each other by a vertical 5G or 6G, also called 5G or 6G infrastructure. The services offered by the boxes depend on the virtual machines that are positioned there, such as distributed databases, geolocation, voice and video, push to talk, blockchain, etc. These femto data centers can be mobile. In this case, they must be battery-powered, which can be simple in an electric vehicle for example, but more complex in a case without simple charging. An optimization of the energy expenditure is necessary by limiting the size of the femto data center and the number of virtual machines to be executed. The internal technology of data centers can be containerization and, for 6G, function-based technologies.

16.3. Cloud Continuum-based networks

The third path that can be followed is the one coming from the Cloud Continuum that can be seen as an intermediate solution between the two previous ones. Virtual machines can be located in the best possible places depending on the many operating parameters. If a virtual machine needs a lot of RAM memory to perform Big Data analytics in memory, it will be located far from the edge. If, on the other hand, an extremely short reaction time is required, the virtual machine should be located next to the sensor or actuator. For example, in automated vehicles, the shortest possible reaction time is required for emergency braking. On the other hand, to find the best route, this action can be done in a Cloud that can be located far from the vehicle.

Figure 16.10 represents this scenario in which five levels are mentioned for data centers ranging from embedded to hyperscale. In this context, the urbanization processes must be strongly developed to calculate the best position for virtual machines, containers or functions at any time. This position can change, requiring a migration of the virtual machine, container or function. Migrations can be performed hot or cold.

The networks that enable Cloud Continuum belong to the different categories of 5G or 6G and Wi-Fi. The private 6G associated with Wi-Fi could become preponderant with the slices of the telecom operators.

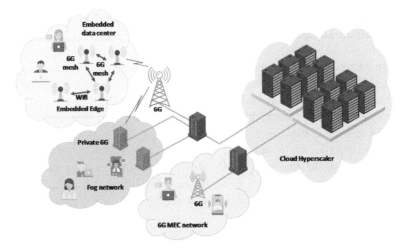

Figure 16.10. *Cloud Continuum-based 6G. For a color version of this figure, see www.iste.co.uk/haddadou/edge.zip*

16.4. Edge and Cloud properties

The future can be summarized in the diagram in Figure 16.11 in which two Edge data centers appear, one of the operator MEC type and the other of the enterprise Fog type. There could also have been embedded Edges. It represents the digital infrastructure of 6G as it can be conceived from 5G. Customer data arrive via a first hop that can use private or public 6G technology, Wi-Fi and then an optical fiber using RoF (Radio over Fiber) technology with signal processing in the data center or EoF (Ethernet over Fiber) with a signal processing in the antenna. All the virtual machines are spread over the data centers following an urbanization software package that has become essential to optimize performance, security, reliability, etc.

Other innovations could intervene in this architecture, such as quantum networks and eventually quantum computers. The advantage of quantum links lies in the immediate detection of eavesdropping: the values of qbits arriving at the receiver are binary and they cannot be those emitted by the transmitter. In the same way, at the level of satellite constellations, many imagine fiber optic networks in the sky or data centers in the satellites to shorten latency times.

Apart from quantum systems and satellites, there will be many advances in nanosecond synchronization of antennas and data centers, centimeter-level locations, and a very strong impact of artificial intelligence.

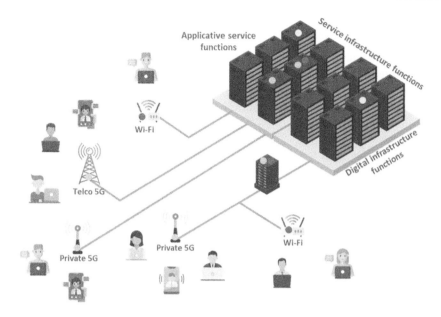

Figure 16.11. *The digital infrastructure of the future. For a color version of this figure, see www.iste.co.uk/haddadou/edge.zip*

Another example is the progression of programming and software development techniques. Figure 16.12 represents this progression toward "zero-code" techniques that will use artificial intelligence programming platforms. In this figure, we can see the transition from the first techniques based on bare metal and then hypervisors to containers and the microservices that find their place in them, and then we see the arrival of functions that are in fact decompositions of microservices to nanoservices. Finally, we will enter the world of low code and no code.

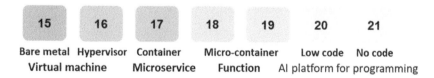

Figure 16.12. *Advanced 6G programming technologies. For a color version of this figure, see www.iste.co.uk/haddadou/edge.zip*

However, the system coding may remain intact and the zero-code may not be used.

16.5. Conclusion

We have studied a rather distant future, but it takes time to develop an entire infrastructure associated, and the transition period between 5G and 6G could be longer than imagined. We can also imagine that 6G will have recovered frequency bands from digital terrestrial television. These bands are very low, and they can be considered as diamond bands.

16.6. References

Abbasi, M., Shahraki, A., Barzegar, H.R., Pahl, C. (2021). Synchronization techniques in device to device- and vehicle to vehicle-enabled cellular networks: A survey. *Computers & Electrical Engineering*, 90, 106955.

Allal, I., Mongazon-Cazavet, B., Al Agha, K., Senouci, S.-M., Gourhant, Y. (2017). A green small cells deployment in 5G – Switch ON/OFF via IoT networks & energy efficient mesh backhauling. *IFIP Networking Conference (IFIP Networking) and Workshops*, Stockholm, Sweden.

Amokrane, A., Zhani, M., Langar, R., Boutaba, B., Pujolle, G. (2013). Greenhead: Virtual data center embedding accross distributed infrastructures. *IEEE Transactions on Cloud Computing*, 1(1), 36–49.

Amokrane, A., Langar, R., Boutaba, R., Pujolle, G. (2014). Energy efficient management framework for multihop TDMA-based wireless networks. *Computer Networks*, 62(7), 29–42.

Amokrane, A., Langar, R., Boutaba, R., Pujolle, G. (2015a). Flow-based management for energy efficient campus networks. *IEEE Transactions on Network and Service Management*, 12(4), 565–579.

Amokrane, A., Langar, R., Zhani, M.F., Boutaba, R., Pujolle, G. (2015b). Greenslater: On satisfying green SLAs in distributed clouds. *IEEE Transactions on Network and Service Management*, 12(3), 363–376.

Avasalcai, C., Murturi, I., Dustdar, S. (2020). *Edge and Fog: A Survey, Use Cases, and Future Challenges*. John Wiley & Sons, Ltd, New York, USA.

Chen, M., Yang, Z., Saad, W., Yin, C., Poor, H.V., Cui, S. (2021). A joint learning and communications framework for federated learning over wireless networks. *IEEE Transactions on Wireless Communications*, 20(1), 269–283.

Chowdhury, M.Z., Hossan, M.T., AIslam, A., Jang, Y.M. (2018). A comparative survey of optical wireless technologies: Architectures and applications. *IEEE Access*, 6(98), 19–40.

Dang, S., Amin, O., Shihada, B., Alouini, M.S. (2020). What should 6G be? *Nature Electronics*, 3(1), 20–29.

Diallo, E., Al Agha, K., Dib, O., Laubé, A., Mohamed-Babou, H. (2020). Toward scalable blockchain for data management in VANETs. In *Artificial Intelligence and Network Applications*, Barolli, L., Amato, F., Moscato, F., Enokido, T., Takizawa, M. (eds). Springer, Cham, Switzerland.

Elkhodr, M. and Hassan, Q. (2017). *Networks of the Future: Architectures, Technologies, and Implementations*. Chapman & Hall, Boca Raton, FL, USA.

Eyassu Dilla Diratie, E.D. and Al Agha, K. (2020). Hybrid Internet of Things network for energy efficient video surveillance. *IEEE 6th World Forum on Internet of Things (WF-IoT)*.

Giordani, M., Polese, M., Mezzavilla, M., Rangan, S., Zorzi, M. (2020). Toward 6G networks: Use cases and technologies. *IEEE Communications Magazine*, 58(3), 55–61.

Gui, G., Liu, M., Tang, F., Kato, N., Adachi, F. (2020). 6G: Opening new horizons for integration of comfort, security, and intelligence. *IEEE Wireless Communications*, 27(5), 126–132.

Gupta, R., Shukla, A., Tanwar, S. (2020). BATS: A blockchain and AI-empowered drone-assisted telesurgery system towards 6G. *IEEE Transactions on Network Science and Engineering*, 8(4), 2958–2967.

Gyongyosi, L. and Imre, S. (2019). A survey on quantum computing technology. *Computer Science Review*, 31, 51–71.

Huang, T., Yang, W., Wu, J., Ma, J., Zhang, X., Zhang, D. (2019). A survey on green 6G network: Architecture and technologies. *IEEE Access*, 7(175), 758–768.

Jiang, W. and Schotten, H.D. (2021). The kick-off of 6G research worldwide: An overview. *Proceedings of 2021 IEEE Seventh International Conference on Computer and Communications (ICCC)*, Chengdu, China.

Jiang, W. and Schotten, H.D. (2022), Initial beamforming for millimeter wave and terahertz communications in 6G mobile systems. *Proceedings of 2022 IEEE Wireless Communications and Networking Conference (WCNC)*, Austin, TX, USA.

Khan, L.U., Yaqoob, I., Imran, M., Han, Z., Hong, C.S. (2020). 6G wireless systems: A vision, architectural elements, and future directions. *IEEE Access*, 8(147), 29–44.

Laubé, A., Martin, S., Quadri, D., Al Agha, K. (2016). Optimal flow aggregation for global energy savings in multi-hop wireless networks. In *Ad-hoc, Mobile, and Wireless Networks. ADHOC-NOW*, Mitton, N., Loscri, V., Mouradian, A. (eds). Springer, Cham, Switzerland.

Laubé, A., Martin, S., Quadri, D., Al Agha, K., Pujolle, G. (2017). FAME: A Flow Aggregation MEtric for shortest path routing algorithms in multi-hop wireless networks. *IEEE Wireless Communications and Networking Conference (WCNC)*.

Laubé, A., Quadri, D., Martin, S., Al Agha, K. (2019a). A simple and efficient way to save energy in multihop wireless networks with flow aggregation. *Journal of Computer Networks and Communications*, 2019, 7059401.

Laubé, A., Martin, S., Al Agha, K. (2019b). A solution to the split & merge problem for blockchain-based applications in ad hoc networks. *8th International Conference on Performance Evaluation and Modeling in Wired and Wireless Networks*, Paris, France.

Letaief, K.B., Chen, W., Shi, Y., Zhang, J., Zhang, Y.-J.A. (2019). The roadmap to 6G: AI empowered wireless networks. *IEEE Communications Magazine*, 57(8), 84–90.

Li, T., Sahu, A.K., Talwalkar, A., Smith, V. (2020a). Federated learning: Challenges, methods, and future directions. *IEEE Signal Processing Magazine*, 37(3), 50–60.

Li, X., Wang, Q., Liu, M., Li, J., Peng, H., Piran J., Li, L. (2020b). Cooperative wireless-powered NOMA relaying for B5G IoT networks with hardware impairments and channel estimation errors. *IEEE Internet of Things Journal*, 8(7), 5453–5467.

Lu, Y. and Zheng, X. (2020). 6G: A survey on technologies, scenarios, challenges, and the related issues. *Journal of Industrial Information Integration*, 19, 100158.

Lv, Z., Qiao, L., You, I. (2020). 6G-enabled network in box for internet of connected vehicles. *IEEE Transactions on Intelligent Transportation Systems*, 22(8), 5275–5282.

Manzalini, A. (2020). Quantum communications in future networks and services. *Quantum Reports*, 2(1), 221–232.

Mohamed, K.S., Alias, M.Y., Roslee M., Raji, Y.M. (2021). Towards green communication in 5G systems: Survey on beamforming concept. *IET Journal*, 15(1), 142–154.

Piran, M. and Suh, D.Y. (2019). Learning-driven wireless communications, towards 6G. *IEEE International Conference on Computing, Electronics & Communication Engineering*.

Qiu, T., Chi, J., Zhou, X., Ning, Z., Atiquzzaman, M., Wu, D.O. (2020). Edge computing in Industrial Internet of Things: Architecture, advances and challenges. *IEEE Communications Surveys Tutorials*, 22(4), 2462–2488.

Rout, S.P. (2020). 6G wireless communication: Its vision, viability, application, requirement, technologies, encounters and research. *11th International Conference on Computing, Communication and Networking Technologies (ICCCNT)*.

Shafin, R., Liu, L., Chandrasekhar, V., Chen, H., Reed, J., Zhang, J.C. (2020). Artificial intelligence-enabled cellular networks: A critical path to beyond-5G and 6G. *IEEE Wireless Communications*, 27(2), 212–217.

Shahraki, A., Taherkordi, A., Haugen, O., Eliassen, F. (2020). Clustering objectives in wireless sensor networks: A survey and research direction analysis. *Comput. Networks*, 180, 107376.

Shahraki, A., Abbasi, M., Jalil Piran, M., Chen, M., Cui, S. (2021). A comprehensive survey on 6G networks: Applications, core services, enabling technologies, and future challenges. *CoRR*, abs/2101.12475.

Shrit, O., Martin, S., Al Agha, K., Pujolle, G. (2017). A new approach to realize drone swarm using ad-hoc network. *16th Annual Mediterranean Ad Hoc Networking Workshop (Med-Hoc-Net)*.

Tariq, F., Khandaker, M.R., Wong, K.K., Imran, M.A., Bennis, M., Debbah, M. (2020). A speculative study on 6G. *IEEE Wireless Communications*, 27(4), 118–125.

Viswanathan, H. and Mogensen, P.E. (2020). Communications in the 6G era. *IEEE Access*, 8(57), 63–74.

Zhang, L., Liang, Y.-C., Niyato, D. (2019). 6G visions: Mobile ultrabroadband, super Internet-of-Things, and artificial intelligence. *China Communications*, 16(8), 1–14.

Conclusion

Edge and Cloud Networking has taken a major place in the networking mode. The developments in the early 2020s come from several points. The first is the Cloud Continuum, which introduces data centers ranging from the infinitely small to the infinitely large. Services must be positioned on this set to enable the associated software to run in the best conditions, taking into account all the criteria of performance, security, availability, resilience, etc. The second point is the adoption of this technique in the context of 5G. This second point is pushing cloud providers to become operators. The three biggest web companies, Amazon, Microsoft and Google, are getting more and more involved in the combination of networks and the Cloud. They lacked the local infrastructure and security. They are establishing themselves by offering local infrastructures (based on private 5G, for example, for AWS), including Edges closer and closer to the users. The third point concerns the extension of networks and the Cloud to storage, computing and security. For security, security Clouds will become very fashionable with a lot of virtual machines, of all kinds.

List of Authors

Kamel HADDADOU
Sorbonne University
Groupe GANDI
Paris
France

Guy PUJOLLE
Sorbonne University
Paris
France

Index

5G, 127–134, 136, 137, 200–203, 205, 207
5GC, 127, 128
6G, 267, 268, 270–278

A, B, C

accelerator, 253, 254, 258, 260
AI, 211, 214, 215, 217, 218, 220
architecture, 31, 32, 36, 40–42, 44–46, 49–51, 53
blockchain, 233, 234, 246
C-V2X, 186–188
Cloud
   -RAN, 130–132
   continuum, 19, 22, 23, 25, 28, 275, 276
   native, 36, 41–45, 47, 54
   Networking, 1, 3, 9, 16, 24, 57, 69, 88, 113, 127, 132, 138, 141, 156, 211, 214, 217, 218, 229, 234–236, 256, 267

D, E, F

data center, 1–3, 5, 7, 9, 12–14, 16, 19–22, 24, 25, 27, 37–39, 45, 47, 49, 52, 57, 73–76, 78, 79, 87–89, 91, 97, 99–102, 106–111, 113, 127–130, 132, 134, 136–138, 141–146, 152, 155, 157, 160, 162, 164, 165, 167, 171, 172, 178, 182, 190, 191, 205, 207, 212, 214, 218, 221, 223, 229, 237, 239–241, 243, 244, 247, 268, 269, 271–276
digital
   infrastructure, 1, 3, 9, 11, 13, 16, 21, 24, 25, 31, 33, 34, 74, 75, 129, 130, 132, 141, 211, 242, 270, 272, 275–277
   twins, 218, 220, 221

digitization of companies, 23, 24
DPDK, 253–257, 259, 263
Edge Networking, 13, 127, 136, 211, 241
fabric, 87–89, 91–94
FD.io, 253, 258, 259, 263

G, H, I

G5, 183, 184, 193
Gaia-X, 49, 52–54
HaLow, 156, 157, 164–166
hardware virtualization, 253, 260, 263
HSM, 229, 231, 232, 245, 246
I2RS, 113, 122, 123
Industry 4.0, 199–207
IoT, 155–158, 162, 165–173, 175, 176, 178

L, M, N

LAN, 162
LISP, 142, 152, 153
management, 211–213, 217
MEC, 127–130, 132, 136, 137, 204, 205, 207
MEF, 146, 147
NB-IoT, 155, 157, 166–169
NFV, 14, 15
northbound interface, 76, 82, 83
NVGRE, 142, 146

O, P, R

OIF, 36, 41, 45
ONAP, 57, 58, 60, 61, 63, 65

ONF, 74–76, 78, 80, 81
Open source, 57, 64, 68, 69
OpenDaylight, 57, 61, 64, 65, 68
OpenFlow, 113–124
OpFlex, 121–123
OPNFV, 57–60, 65
P4, 113, 120, 121, 124
PAN, 162
RoF, 141, 143

S, T, V

SASE, 102, 103, 105, 106
SD-WAN, 87, 96–103, 105, 106
SDN, 73, 74, 77–79, 84
security, 229–246
SIM, 229–231, 245
southbound interface, 76, 78, 80
TRILL, 141, 150, 151
vCPE, 87, 101, 102, 105, 106
vehicular networks, 181, 185, 189, 190, 192, 193
virtualization, 1, 2, 8, 9, 11, 12, 14, 15, 31, 57–59, 65, 73, 76, 77, 80, 87, 89, 91, 95, 105, 107–111, 127, 130, 132, 134, 137, 143, 146, 183, 203, 204, 235, 236, 241–243, 253, 254, 256, 260, 263, 268
VLC, 182
vRAN, 87, 109, 110
VxLAN, 142, 144, 145

Printed and bound by CPI Group (UK) Ltd, Croydon, CR0 4YY
20/12/2023